Brock/Springer Series in Contemporary Bioscience

# Ecology of
# Protozoa

# Brock/Springer Series in Contemporary Bioscience

*Series Editor: Thomas D. Brock*
*University of Wisconsin-Madison*

ECOLOGY OF PROTOZOA: The Biology of Free-living Phagotrophic Protists
  *By Tom Fenchel*

**Forthcoming in this series:**

ROOT-ASSOCIATED NITROGEN-FIXING BACTERIA
  *By Johanna Döbereiner*
  *Fabio O. Pedrosa*

Tom Fenchel

# Ecology of Protozoa

## The Biology of Free-living Phagotrophic Protists

With 47 Figures

**Science Tech Publishers**
Madison, Wisconsin

**Springer-Verlag**
Berlin   Heidelberg   New York
London   Paris   Tokyo

Prof. Tom Fenchel
Department of Ecology and Genetics
University of Aarhus
DK-8000 Aarhus
Denmark

**Library of Congress Cataloging-in-Publication Data**

Fenchel, Tom.
　Ecology of Protozoa.

　(Brock/Springer series in contemporary bioscience)
　Bibliography: p.
　Includes index.
　1. Protozoa—Ecology. I. Title. II. Series.
QL366.F46　1987　　593.1′045　　86-20300
ISBN 0-910239-06-1

Science Tech, Inc., 701 Ridge Street
Madison, Wisconsin 53705, U.S.A.

Sole distribution rights outside of the USA, Canada, and Mexico granted to
Springer-Verlag Berlin Heidelberg New York London Paris Tokyo

ISBN 0-910239-06-1 Science Tech, Inc. Madison, WI
ISBN 3-540-16960-1 Springer-Verlag Berlin Heidelberg New York

Production and editorial supervision: Science Tech Publishers
Interior design: Thomas D. Brock
Cover design: Katherine M. Brock
Media conversion and typesetting: Impressions, Inc., Madison, Wisconsin
Printing: Braun-Brumfield, Ann Arbor, Michigan

Printed in the United States of America
10 9 8 7 6 5 4 3 2 1

# Preface

This book is written for ecologists and protozoologists. Ecologists who study environments and biotic communities in which protozoa are important should find this book especially useful. During the last decade it has become clear that protozoa play important roles in natural ecosystems, but few ecologists have a feeling for the functional properties and the diversity of these organisms. Protozoa pose or exemplify many general problems of population and community ecology, and of evolutionary biology. In most respects the general ecological properties of protozoa are not fundamentally different from those of larger organisms; yet, due to their small size, short generation times, and ubiquitous occurrence they often present ecological phenomena in a new and different light. To this should be added that protozoa are well-suited for experimental work. Despite these advantages, the study of protozoa has played a relatively modest role in the development of ecology and evolutionary biology, primarily, I believe, because most ecologists are unfamiliar with these organisms. I hope this book will attract more attention to these favorable characteristics of protozoa.

I also hope that this book may make protozoologists aware of new aspects of their pet organisms. For a long time (that is, until the fundamental distinction between prokaryotic and eukaryotic cells was recognized) protozoa were believed to represent the simplest form of life. They were therefore extensively used for the experimental study of basic questions of cell biology. Protozoa (and not least the ciliate *Tetrahymena*) are still used in the laboratory as cell models, but prokaryotes and plant and animal tissue cultures have to a large extent taken over this role. Protozoa are eukaryotic cells which are exposed directly to the environment and they are subject to natural selection as individual organisms. Protozoan cells therefore show a variety of specialization and complexity of structure and function which are unchallenged among other types of eukaryotic cells. Cell biology may therefore still harvest much from the study of protozoan cells, not because of their "primitive" or "generalized" nature, but because of their specializations, which permit the study of certain features of cells which are somehow amplified in a particular species. However, in order to fully exploit this aspect of protozoa, it is necessary to understand the relationships between the organisms and their environments and the adaptive nature of the studied traits. Throughout this book I emphasize the role of physiological and structural constraints for understanding the role of organisms in nature and the close relationship between cell physiology and ecological insight.

This book is divided into twelve chapters. The first one gives a general introduction to the nature of unicellular eukaryotic organisms in general and to protozoa (defined functionally as phagotrophic, unicellular organisms) in particular. The following five chapters are devoted to the functional biology of protozoa: how they move and orient themselves in the environment, their bioenergetics, symbiotic relationships with other organisms, etc. These chapters serve as a necessary background for understanding protozoa in their natural environments. Chapter 7 treats general ecological principles (dynamics of food chains, environmental patchiness, niche diversification and biogeography) with special reference to protozoan populations and communities. Together with the physiological properties of protozoa, the considerations offered in this chapter explain properties of protozoan ecology which are common to all types of ecological systems. The following four chapters discuss protozoan communities belonging to different types of habitats. The treatment emphasizes the role of environmental patchiness in time and space, food resource specialization, and protozoan diversity, as well as the role of protozoa in food chains and in the flow of carbon and other elements in a particular ecosystem. The book is concluded by a chapter which offers some general considerations.

I believe that the publication of this book is worthwhile. Although protozoan diversity has recently been monographed (Lee et al., 1985) and many aspects of protozoan physiology and biochemistry have been reviewed by Levandowsky & Hutner (1979–81), there is no general treatment of the ecology of protozoa. I have tried to write a book which will be easy to read and which will inspire the readers with new ideas (rather than a reference book or an annotated bibliography). Consequently, I have chosen to discuss some examples in more detail but I have not attempted to be comprehensive. I apologize to colleagues who may feel that their work is inadequately cited, although it may be as

interesting and illuminating as the examples I do discuss. A recent comprehensive bibliography on the ecology of free-living protozoa is found
in Finlay & Ochsenbein-Gattlen (1982). I have not treated methodology
in any detail, but the reference list should be helpful. The section on
protozoa in Burns & Slater (1982) treats methods of collecting, enumerating, culturing, and identifying protozoa and gives additional
references.

**Acknowledgements**    I am grateful to Dr. Thomas D. Brock who
suggested that I write this book; it was something I had wished to do
for some time, but clearly I needed an external stimulus. Several of the
micrographs (primarily the best ones) have been made by various col-
leagues to whom I express my gratitude for allowing me to use their
micrographs: Dr. Hilda Canter-Lund (Figures 9.1,A; 9.2,A-D; 9.5,A); Dr.
Bland J. Finlay (Figures 2.3,C; 9.1,B,C; 9.6); both of the Freshwater
Biological Association, Windermere Laboratory, United Kingdom; Dr.
Barbara Grimes, Department of Zoology, University of North Carolina,
Raleigh (Figure 5.1); and Helene Munk Sørensen, M.Sc., Ringkøbing
Amtsvandvæsen, Denmark (Figure 8.4,A-K). I am also grateful to Ms.
Ilse Duun Jensen and Ms. Annie Jensen for their assistance with pho-
tographic work. Above all I am grateful to my friends and colleagues
Bland J. Finlay (Windermere Laboratory) and David J. Patterson (De-
partment of Zoology, University of Bristol) for reading the entire man-
uscript; their comments, critical remarks, and suggestions have, I feel,
greatly improved the form and contents of the book.


                                                          TOM FENCHEL

# Contents

Preface ................................................................. v

## 1 What is a Protozoan?    1

1.1 Historical Views on the Nature of Protozoa ........................ 1
1.2 The Origin and Diversification of Protists ........................ 3
1.3 Unicellularity, Death, and Sex .................................... 5
1.4 The Species Concept in Protozoa ................................... 7
1.5 The Structural Complexity of Protozoa ............................. 9
1.6 The Size Range of Protozoa ....................................... 11

## 2 Ecological Physiology: Motility    15

2.1 Introduction .................................................... 15
2.2 How Protozoa Move: Life in Syrup ................................. 16
2.3 Orientation in the Environment ................................... 24

## 3 Ecological Physiology: Feeding    32

3.1 General Considerations ........................................... 32
3.2 Feeding in True Protozoa ......................................... 41

## 4 Ecological Physiology: Bioenergetics    53

4.1 Balanced Growth: The Efficiency of Conversion ................... 53
4.2 Balanced Growth: The Rate of Living ............................. 56
4.3 Nonbalanced Growth: a Feast and Famine Existence ........ 59
4.4 Anaerobic Metabolism ............................................ 61

## 5 Ecological Physiology: Other Aspects    63

5.1 Polymorphic Life Cycles .................................................. 63
5.2 The Adaptive Significance of Sexual Processes ................... 68
5.3 The Physical and Chemical Environment ......................... 72

## 6 Symbiosis    76

6.1 The Definition of Symbiosis ............................................ 76
6.2 Associations with Photosynthetic Organisms .................... 77
6.3 Nonphotosynthetic Symbionts .......................................... 83

## 7 The Niches of Protozoa    86

7.1 Introduction .................................................................. 86
7.2 Steady-State Phagotrophic Food Chains ........................... 87
7.3 Patchiness and Successional Patterns .............................. 90
7.4 Niche Differentiation and Coexistence ............................. 92
7.5 Biogeography of Protozoa ............................................... 97

## 8 Protozoan Communities: Marine Habitats    102

8.1 Introduction .................................................................. 102
8.2 Marine Pelagic Protozoa ................................................. 103
8.3 Marine Sediments .......................................................... 116

## 9 Protozoan Communities: Freshwater Habitats    134

9.1 Differences from Marine Communities ............................. 134
9.2 Pelagic Protozoa ............................................................ 136
9.3 Sediments, Detritus, and Solid Surfaces .......................... 141
9.4 Running Waters ............................................................. 145
9.5 Organic Enrichment and Polluted Waters ........................ 149

## 10 Protozoan Communities: Terrestrial Habitats    152

10.1 The Nature of Terrestrial Protozoa ................................ 152
10.2 The Role of Protozoa in Soil Ecosystems ....................... 157

## 11 Symbiotic Protozoa    161

## 12 Concluding Remarks    167

References    171

Index    195

# 1

# What is a Protozoan?

## 1.1 Historical views on the nature of protozoa

The question as stated by the chapter heading is far from trivial. Many standard textbooks define protozoa as "unicellular animals," but this is not entirely satisfactory. The idea that a protozoan is unicellular in the sense that it corresponds to a single cell of a multicellular organism, was first conceived and accepted about 170 years after the discovery of protozoa by Leeuwenhoek in 1674. The term "Protozoa" was coined by Goldfuss in 1817 to mean "original animals" and it included the coelenterates. The title of Ehrenberg's memoir: *Die Infusionsthierchen als vollkommene Organismen* (1838), which also included small multicellular creatures such as rotifers, alluded to the idea that protozoa were quite comparable to higher animals, the feeding vacuole to a stomach, and so on. D'Orbigny, who coined the name Foraminifera in 1826, considered them to be a type of cephalopod because of the resemblance between the test of (some) foraminifera and the shell of the mollusc *Nautilus*. It was only in the middle of the last century, after the cell was recognized as a building unit of animals and plants, that the idea of protozoa as single cells comparable to cells of multicellular organisms became generally accepted. This idea then led to the theory that multicellular organisms originated as protozoan cell colonies.

In a certain sense, Ehrenberg's view on the nature of protozoa enjoyed a mildly successful revival in this century. The way we view unicellular eukaryotes is, of course, closely linked to our view of the origin of multicellular beings. Thus it has been suggested that the most primitive animals are turbellarians and that these originated not as a protozoan

1

colony, but rather through compartmentalization of a ciliated protozoan which then became "acellular" rather than unicellular (see e.g., Hanson, 1976). As a theory of phylogeny this has little appeal. However, it is mentioned here because, in one sense, it has been adopted in this book. From an ecological and evolutionary point of view, a protozoan cell is a unit, comparable to an individual animal rather than to one of its constituent cells. Protozoa interact with their environment and with other organisms just as metazoa do, and the individual, whether a unicellular or a multicellular organism, is the basic unit of natural selection. Thus, in this sense, protozoa are "complete organisms" in the meaning of Ehrenberg.

Biologists have always classified the living world into two basic "kingdoms"; that of animals and that of plants (although the botanists have slowly yielded the fungi and the prokaryotes to mycologists and microbiologists, respectively). In this tradition, the unicellular eukaryotes were shared among zoologists, botanists, and mycologists. The most "animal-like" were called protozoa and adopted by zoologists. These organisms included mainly the motile, phagocytotic forms without photosynthetic pigments. The botanists took over the "protophytes" which usually possess chlorophylls and, finally, some fungi-like amoebae were treated together with the real fungi in taxonomic contexts. This rather arbitrary classification resulted in certain groups (e.g., euglenids and dinoflagellates) being claimed by both botanists and zoologists, while the myxamoebae appeared in zoology as well as mycology textbooks. It was also recognized that many forms display animal as well as plant-like features and that among the flagellates, some very closely related photosynthetic and non-photosynthetic species exist. In fact, the distinction between protozoa, protophytes, and "lower fungi" no longer agrees with our views on evolutionary relationships. The term "Protista," originally coined by Haeckel (1866), has been revived to include a large number of eukaryotes, often quite unrelated, which are neither animals (namely, metazoa), vascular plants, or higher fungi (Corliss, 1984). Some of these eukaryotic lineages have given rise to plants, animals, and fungi; in general, the origin of and the relationships between these lineages is incompletely understood. In many cases their origin probably began in the Precambrian period, shortly after the origin of the first eukaryotic organism.

The concept of Protozoa has no particular evolutionary or systematic meaning. Since we are primarily concerned with ecology, we can define protozoa in a functional sense as phagotrophic protists. Furthermore, since we will only consider free-living forms, we will include the ciliates (a monophyletic, "natural" group), the flagellates (or most of them), and the sarcodines. The two latter groups each consist of a number of largely unrelated evolutionary lineages. Some groups of organisms which are traditionally included in protozoology texts will receive little or no attention. These are obligatory parasites (the four groups previously assembled under the name Sporozoa, a few flagellate groups, and some minor ciliate taxa) or they are exclusively photoautotrophs or "saprozoic" (feeding on dissolved organic material) and at the same time are incapable of phagocytosis (some "phytoflagellate" groups and some fungi-like sarcodines).

## 1.2   The origin and diversification of protists

The most fundamental distinction which can be made in the living world is that between prokaryotes and eukaryotes. The protists (and hence the protozoa) unambiguously belong to the eukaryotes. They possess a nuclear envelope, eukaryotic ribosomal RNA, and endoplasmic membranes. They have characteristic eukaryotic organelles (mitochondria, chloroplasts, and flagella), histones associated with their chromosomal DNA, and the ability to perform phagocytosis—although one or more of the above mentioned features may be absent in some forms. Also, eukaryotic cells (and thus protists) are in almost all cases larger than prokaryotic cells.

In spite of the apparent discontinuity between prokaryotes and eukaryotes, it is generally assumed that the latter are descended from the former in some way. This is because eukaryotes have a much more complex structure and since it is now known from the geological record that prokaryotes came into existence perhaps two billion years before the origin of the eukaryotes.

The most accepted explanation for the gap in structural complexity between prokaryotes and eukaryotes is that the latter are hybrid organisms: That the original "pro-eukaryote," which was unique because it had the ability to phagocytize, acquired prokaryotic endosymbionts which originally entered the cells as prey or parasites. Eventually, these developed into cellular organelles. Evidence for this theory of "serial endosymbiosis" has been accumulated by Margulis (1981). The evidence is particularly convincing as regards the origin of chloroplasts and of mitochondria. It explains the presence of an organelle genome, bacterial type ribosomes in the organelles, and homologies in the amino acid sequences between proteins of organelles and extant prokaryotes. Circumstantial evidence comes from various types of endosymbiotic relationships of extant protists (discussed in some detail in Chapter 6). This theory implies that the ability to perform phagocytosis was an original feature of the eukaryote ancestor which was then a protozoan in the sense of this book. The evidence also suggests that some eukaryote features evolved independently in different evolutionary lineages. This almost certainly applies to the acquisition of chloroplasts, which apparently evolved from different types of cyanobacteria in different photosynthetic protists, most of which are unknown or extinct. It must be added, however, that there are aspects of eukaryote structure which are not explained by or easily reconciled with the theory of serial endo-symbiosis as presented by Margulis (1981). These include the origin of flagella, basal granules and other microtubular structures, and the origin of mitosis and of sexuality.

Indirect evidence in the form of molecular genealogy (nucleotide and amino acid sequences) suggests that the eukaryotes, and therefore the protists, originated and diversified around $1.5 \times 10^9$ years ago (Margulis, 1981; Nanney, 1982). Direct palaeontological evidence is scarce. Only cells with rigid walls, scales, cysts, tests, or loricas can be expected to leave any recognizable fossil remains. The earliest convincing remains of eukaryote cells are about $850 \times 10^6$ years old. In 700 to $800 \times 10^6$ year old deposits from many parts of the world, so called "chitinozoa" are usually interpreted as fossil protozoa (Glaessner, 1984, and references therein). Molecular genealogy and the relatively large variation in cytological detail among protists as compared to that found between different types of metazoa suggest that the main groups of extant protists had already diverged long before Phanerozoic times. The ciliates, in particular, seem to demonstrate this; in addition to a large number of peculiar features (nuclear dimorphism, cortical organization), there is evidence of differences in the genetic code relative to other organisms (Caron & Meyer, 1985).

From Phanerozoic times there is a considerable amount of evidence from protozoan forms which produced tests, cell walls, or skeletal material. This supports the idea that most major groups of protozoa had arisen well before the Cambrian period. The two main groups of radiolaria are represented in deposits from early Palaeozoic times, as are tintinnids (specialized planktonic ciliates which build loricas) and foraminifera. Some other groups, e.g., the dinoflagellates, seem to appear only later in the fossil record (Anderson, 1983; Lee et al., 1985). There are other groups of protozoa which must have evolved during more recent geological time. For example, the highly specialized entodiniomorphid ciliates occur only in the digestive tract of herbivorous mammals, in particular, the ruminants. Clearly, these forms must have evolved from some free-living ancestor during Cenozoic times and perhaps as late as the Miocene, when ruminant mammals appeared.

We still have only a very incomplete understanding of the phylogenetic relationships between the major groups (phyla) of protists (of which Corliss, 1984, recognizes 45). This also applies to the more restricted number of phyla included in the protozoa in the sense of this book. There is no doubt, however, that this situation is likely to improve in the future when a large amount of data on nucleotide and amino acid sequences from many types of protists will be available.

## 1.3   Unicellularity, death, and sex

The definition of protozoa as unicellular organisms is not as simple as it might sound. If we discuss all protists we must obviously also include multicellular organisms such as large algae, but even among the protozoa we find many examples of multicellularity, which have arisen independently. Many large protozoa (e.g., larger ciliates, amoebae, and foraminifera) are multinucleate. Functionally, no doubt, this reflects the need for a large nuclear surface area through which enough mRNA can be transported in order to supply a large cell. This situation may have been brought about by endomitosis or, in some cases, by fusion of single cells into a syncytium (in the Eumycetozoa, the acellular slime molds). Other species may form cell colonies which obviously originated from incomplete cell divisions (e.g., *Polykrikos* and colonial peritrichs, see Figure 1.1). In all cases, however, these colonial forms seem to lack one distinctive feature of metazoa and vascular plants: they do not form tissues of specialized cells.

There would seem to be another fundamental difference between true multicellular organisms and protozoa. An individual metazoan develops from a fertilized egg or (in the case of parthenogenesis) an unfertilized

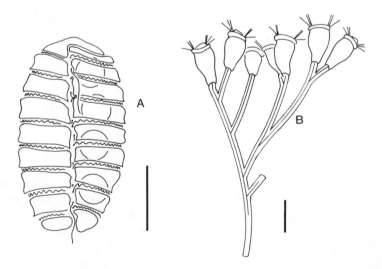

**Figure 1.1**   Colonial protozoa. A. *Polykrikos schwartzi*, a phagotrophic dinoflagellate which may have evolved through incomplete cell divisions in an ancestor which looked like a typical dinoflagellate. It has four nuclei but eight pairs of flagella; typical dinoflagellates have only one pair of flagella and one nucleus (redrawn from Grassé, 1952). B. *Zoothamnium hiketes*, a colonial peritrich found on the body surface of the amphipod *Gammarus*. Each individual colony develops from a single cell through repeated cell divisions. (Redrawn from Fenchel, 1965.) Scale bars for A and B: 50 μm.

egg. The resulting phenotype is a cell colony which is the result of mitotic divisions of the egg cell and which determines the Darwinian fitness of the individual (that is, the number of offspring it will produce), but it is mortal. Only the gonadal cells are potentially immortal; the somatic part of the phenotype is doomed. A protozoan (at least an asexual species, see below) does not live with this regrettable fact; it is potentially immortal and may in theory go on living and dividing forever.

However, the tendency to evolve these multicellular traits (namely, cell specialization within colonies and "built-in" death) can also be found among protozoa. The best example of cell tissue among protozoa is provided by the Myxosporidia (which includes only obligatory parasites). Here the spore, a dispersive stage, consists of several specialized cell types, of which only one develops in the new host individual. Death is also a property of colonial forms. The best-known one is *Volvox*, a photosynthetic, colonial flagellate which is not really a protozoan in the present sense. It forms spherical colonies and new colonies develop from generative cells inside the mother colony. The young colonies can escape and start their independent life only after the mother colony breaks open and succumbs: Hence inevitable death was introduced in the protistan world. (This is beautifully illustrated in a cartoon in the book by Hegner, 1938, which shows young *Volvox* colonies mourning at the wake of their burst mother.) Many sessile peritrich ciliates also form colonies. In some of the larger ones (e.g., *Zoothamnium geniculatum*, whose colonies may grow to several millimeters in height) there are two kinds of cells. One kind is capable of dividing to form new sessile members of the colony, but it cannot form the free-swimming cells with a "telotroch" (a ciliary girdle around the equator of the cells) which can establish new colonies. Each colony seems to have a restricted life span (Wesenberg-Lund, 1925).

This tendency to develop "metazoan characteristics" has occurred independently among many protistan lineages and lends credence to the theory that the higher animals and plants evolved from colonial protists. Biochemical and structural evidence makes it highly probable that the vascular plants evolved via the green algae from organisms like *Volvox*. There is some similar evidence that colonial choanoflagellates (which are still extant) gave rise to the sponges and perhaps to other metazoans as well.

Members of many groups of protozoa apparently do not have sexual processes. This applies to most amoebae, to several flagellate groups, and to several minor taxa within most major groups of protozoa (the distribution of sexuality and its adaptive nature is discussed in more detail in Section 5.2). Other groups of protozoa include organisms which have sexual processes as an obligatory part of their life cycle or under certain environmental conditions. In most cases (and in contrast to the beliefs of many puritans) the protozoa demonstrate that sex is not necessarily coupled to reproduction. In ciliated protozoa, two cells attach to each other and exchange haploid micronuclei (conjugation). In this type of sexuality a "migratory" nucleus fuses with a "stationary" nucleus within each conjugant cell to form a diploid syncaryon. Another type of ciliate sexuality (autogamy) occurs when two haploid micronuclei deriving from the same micronucleus fuse within an individual cell. In both cases this leads to a genetic reshuffling (complete homozygosity in the case of autogamy) and the formation of a new macronucleus from the new syncaryon. But the process is not coupled to reproduction, which takes place in subsequent cell divisions during which the micronucleus undergoes an endomitosis and the polyploid macronucleus divides amitotically. (In some ciliates, the Karyorelectida, the diploid macronucleus cannot divide, but is replaced through mitosis of a micronucleus during cell fission.) Ciliates which have sex in a sense are analogous to metazoa which are comprised of somatic, mortal parts representing the phenotype (the somatic cells in metazoa, the macronucleus in ciliates), and potentially immortal generative parts (the gonadal cells in metazoa, the micronucleus in ciliates). Some ciliates, however, do not have sex and some are even devoid of a micronucleus.

Other protozoa which have sex produce gametes which fuse to form other gametes. Meiosis may be gametic or zygotic according to whether the major part of the life cycle is spent in a haploid or a diploid state. Again, sexuality in most cases is not coupled to cell multiplication. In some forms, such as the foraminifera, however, sexuality is a fixed part of the life cycle and takes place as a mass release of gametes and so also functions as a reproductive process.

## 1.4 The species concept in protozoa

The ability to recognize species and the information which is built into the taxonomic system are both vital components of field and experimental ecology. Without the confident identification of organisms, the interpretation of experimental results and observations would become imprecise and possibly not reproducible. Unfortunately, the species systematics of many protozoan groups still poses problems and its degree of resolution is probably quite uneven.

In common with other eukaryotic groups of organisms, protozoan systematics is based on morphological traits. However, the amount of detail which can be observed in protozoa varies tremendously from group to group. The fact that the two protozoan groups in which most species have been described are the foraminifera and the radiolaria is probably not accidental. Both groups include large organisms which produce skeletal material or tests which can easily be preserved and which reveal a considerable amount of detail in the light microscope. Conversely, small amoebae show comparatively little detail and the species systematics is proportionately cruder. Some protozoa are very small and the amount of detail seen in the light microscope is insufficient as the basis of species description and sometimes insufficient even for assigning them to a higher taxon. This applies in particular to many heterotrophic flagellates, which are ecologically important in soils, as well as in aquatic environments. A satisfactory species taxonomy based on morphological traits requires electron microscopy, but the methodical application of this technique has hardly begun.

Ciliates are generally larger organisms and they show a considerable amount of detail on their cell surfaces with regards to patterns of cilia, kinetosomes, and fibrillar structures. The development of permanent staining methods which show these structures has improved ciliate systematics and led to the establishment of a type-specimen collection (Corliss, 1972), a new concept in protozoan taxonomy. Of particular value is the establishment of culture collections (Hutner, 1975; Lee et al., 1985) which are indispensable when working with sibling species or when biochemical criteria are used.

With regard to the concept of "genetic" or "evolutionary" species, different and biologically interesting problems arise. Many protozoa seem to be entirely sexless or obligatory selfers. In such forms, a biological species concept does not really exist. One might say that each clone constitutes its own species. This does not mean that sexless forms do not often occur as discrete groups with characteristic phenotypes, and naming them is both possible and desirable. But this exercise has no evolutionary meaning in the sense of populations of organisms sharing a common gene pool.

In outbreeding forms it is possible, at least in principle, to establish biological species. Such an analysis was first attempted by Sonneborn on *Paramecium aurelia* (see Sonneborn, 1957, 1975). Mating experiments showed that this "species" is really a complex of fourteen sibling species which are genetically isolated. Closer studies have revealed that these sibling species (or syngenes as Sonneborn called them) do in fact show slight morphological differences and some of them may be identified microscopically. They also show ecological differences with respect to their habitat preference and life cycle characteristics, and they show a differential geographic distribution; some of them only occur on certain continents.

A similar system has been established by Nanney and co-workers (Nanney, 1982; Nanney et al., 1980) for the popular ciliate, *Tetrahymena pyriformis*. So far, this complex has been shown to consist of seventeen isolated groups. Among these, however, there are four nonsexual, amicronucleate strains characterized only by their isozyme patterns. The others are outbreeding, true species. As in the case of the *Paramecium aurelia* complex, the different species show some zoogeographical differences. They are, however, practically impossible to distinguish. It was therefore surprising to find that with respect to isozyme patterns, as well as in amino acid and nucleotide sequences of other macromolecules, they show substantial differences and a variation which far exceeds that found among mammals, for example. Also, the amicronucleate strains differed considerably from any of the sexual strains. It was concluded that these species (including the asexual ones) have had a long independent evolution perhaps spanning the entire Phanerozoic period. However, more recent estimates of the age of the *Tetrahymena pyriformis* complex, based on the structure of ribosomal RNA, are only 30 to 40 million years (Nanney, 1985). The reason why these organisms are morphologically so alike must be that their particular morphological phenotype represents an adaptive peak (that is, small deviations will lead to a decrease in fitness so that natural selection has constrained any changes in the morphology of these genetically divergent forms).

With respect to these two examples, it has recently been advocated that the sibling species should be considered as "real species" and named according to normal zoological practice (Corliss & Dagett, 1983). However, this phenomenon—that a species defined on the basis of morphological characters consists in fact of a complex of species with very similar appearances—is probably widespread among ciliates and it is highly unlikely that they will all be unravelled. From the viewpoint of an ecologist the existence of many sibling species poses a number of interesting problems concerning possible ecological differences and how the many sibling species are maintained in nature in spite of exploitative competition from very similar forms.

## 1.5 The structural complexity of protozoa

The concept of protozoa as single-celled organisms gives many people the impression of primitive, archaic organisms, the paradigm of which is the seemingly structureless slime which makes up an amoeba. It is true that the constraints on size at the protozoan level of organization set certain limits on structural complexity. Yet, protozoa have evolved over a time span which exceeds that of metazoan evolution by hundreds of millions of years. During this time span, different kinds of protozoa have adapted to all types of environments and to different prey, predators, competitors, and symbionts. As a result of this, protozoa show specializations, complex morphological adaptations, and life cycles which have no counterpart in the individual cells constituting multicellular organisms.

The external morphology of the hypotrich ciliate *Euplotes* exemplifies this complexity. This is one of the most complex of ciliates and

**Figure 1.2**   The complex surface structures of the hypotrich ciliate *Euplotes moebiusi* as seen with the scanning electron microscope. The ventral side (left) shows bundles of cilia (cirri) with which the cell walks on solid surfaces and (on the left side of the cell) the peristome with the ciliary membranelles. The dorsal side (right) shows the short "sensory" cilia. Scale bar: 10 μm.

Figure 1.2 shows the cirri (bundles of individual cilia) with which the organism can walk on surfaces and shows the membranelles of the mouth with which it filters suspended food particles. Also seen are the dorsal "bristles," short, immobile cilia which, it has been speculated, serve as mechanoreceptors. Figure 1.3,A shows the "Müller vesicle" of the ciliate *Loxodes*. This organelle serves as a gravity sensor, informing the geotactic ciliate which is up and which is down (Fenchel & Finlay, 1984, 1986a). The dinoflagellate *Erythropsis* has a structure called the ocellus (Figure 1.3,B), which in all respects resembles an eye, including a pigment capsule and a lens. No experimental study of this interesting creature exists, unfortunately, but it is usually assumed that the organelle permits phototaxic behavior in this oceanic planktonic organism (Greuet, 1968). A final example of protozoan complexity is the lorica of acanthocoeid choanoflagellates (Figure 8.2,E, here represented by *Diaphanoeca grandis*). These organisms which abound in marine plankton (together with nonloricate choanoflagellates) build complex siliceous loricas which vary considerably in size and structure among

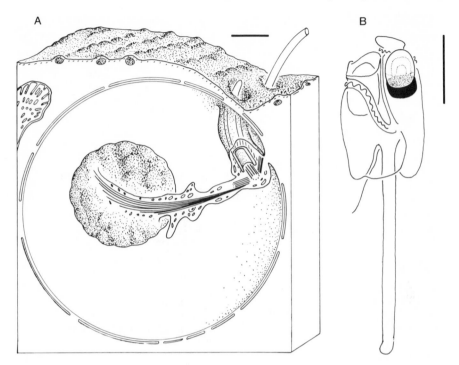

**Figure 1.3**   Complex structures in protozoa. A. The "Müller vesicle" of the ciliate *Loxodes striatus*, an organelle which functions as a gravity sensor. The statolith, (a membrane-covered conglomerate of barite granules) is held by a rigid stalk connected to a pair of basal granules which are invaginated from the cell surface. Scale bar: 1 μm (after Fenchel & Finlay, 1986a). B. The planktonic dinoflagellate, *Erythropsis pavillardi*, with an "eye" consisting of a lens and a pigmented cup. Scale bar: 50 μm. (Redrawn from Greuet, 1969.)

the different species (Fenchel, 1986b). The function of the lorica has so far only been a matter of speculation.

## 1.6   The size range of protozoa

Protozoa have to be studied under the microscope and so they are "microbes." The fact that protozoa are small is of paramount importance for understanding their functional biology and ecological role in nature; this aspect is explored in the following chapters. What is rarely appreciated is the enormous variation in size displayed by protozoa. Figure 1.4 shows a logarithmic scale of length covering a range from $10^{-8}$ cm (the diameter of a hydrogen atom) to more than $2 \times 10^4$ cm (the length of a blue whale) and hence everything of interest to a biologist. It is immediately seen that the size range covered by protozoa (about four

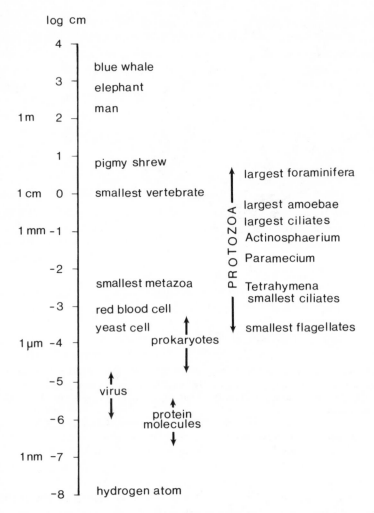

**Figure 1.4** Scales showing the sizes of living things and of protozoa.

orders of magnitude) corresponds to about one-third of the entire scale. If the largest protozoa had exactly the same shape as the smallest ones (which is not the case; large protozoa tend to be flattened or very oblong whereas small ones are nearly spherical) the cell volumes of protozoa would span a range of $10^{12}$. The range of lengths of protozoa exceeds that displayed by vertebrates; the ciliates alone represent a size range comparable to that of all terrestrial mammals (pigmy shrew to elephant). These observations already suggest an enormous diversity in the functional roles of protozoa in nature.

Difference in size among protozoa is synonymous with a difference in individual cell size. This is in contrast to multicellular organisms where body size is primarily a function of the number of cells. A histological section of some tissue from an elephant is virtually indistinguishable from a section of the same tissue from a mouse except that the latter will have a somewhat higher density of mitochondria (Schmidt-Nielsen, 1984). In protozoa, size differences are reflected by a number of adaptations at the cellular level. Thus the numbers of various organelles, such as nuclei, mitochondria, and kinetosomes, increase with increasing cell size and there are obvious constraints on the shapes of very small as well as of very large protozoa.

The smallest eukaryotic cells (certain yeasts, photosynthetic flagellates, and some parasitic protozoa) measure around 2 μm and the smallest free-living, phagotrophic protozoa (chrysomonad flagellates and choanoflagellates) are about 3 μm in diameter. They just overlap in size with the largest prokaryotic cells such as cyanobacteria. It is conceivable that this represents the minimum size of a eukaryotic cell which usually must contain at least one mitochondrion and a nucleus with 100 to 1000 times more DNA than a prokaryotic cell. The ability to perform phagocytosis presumably also requires space for the machinery responsible for motility and a eukaryotic flagellum is more bulky than a bacterial flagellum.

The largest protozoa are represented by certain foraminifera and some sarcodines which may measure up to a centimeter or so. Certain extinct foraminifera were as large as 10 cm (Lee et al., 1985) as are some extant deep-sea sarcodines, the Xenophyophoria (see Figures 6.1,B and 8.9,D). (We do not here consider certain colonial forms of radiolaria, for example, which also measure several centimeters, see Anderson, 1983.) Giant protozoa are all very thin and flat or highly vacuolated. Furthermore, the foraminifera in question seem in all cases to harbor photosynthetic endosymbionts, which in part exclude them from some of the considerations offered below. Certain amoebae, such as *Chaos* and *Pelomyxa*, may measure several millimeters. The largest swimming protozoa are giant ciliates, such as *Stentor* and *Spirostomum*; such large ciliates are always either oblong and worm-shaped, or in some cases very flat and leaf-like.

The basic constraint on the maximum size of a unicellular organism is the rate of diffusion. Within an aerobic cell, oxygen is transported solely by molecular diffusion: therefore, we may calculate the maximum size of a spherical cell for which the oxygen consumption does not exceed the rate at which it can be transported through the cell, namely, the size at which the center of the cell has an $O_2$-concentration of zero.

For a concentration distribution which is spherically symmetrical, the diffusion equation takes the form: $dC/dt = D(r^{-2} d(r^2 dC/dr)/dr)$, in which C is the $O_2$-concentration, r is the radius, and D is the diffusion coefficient (Rubinow, 1975). For our problem we will have to subtract the oxygen consumption rate, R, from the right hand of the above equation. At steady state, the derivative vanishes ($dC/dt = 0$). Thus, integrating the equation, and setting $C(r = 0) = 0$, we find that the maximum radius above which the center of the cell becomes anoxic is given by: $r = (C_o \times 6D/R)^{1/2}$, where $C_o$ is the external $O_2$-concentration. Assuming that $C_o$ is $8 \times 10^{-3}$ ml $O_2$ per ml water (that is, about atmospheric saturation) and that D for oxygen in the cytoplasm is two-thirds of that in water (which is about $2.4 \times 10^{-5}$ cm²/sec) and that the volume specific respiration of a large protozoan is $1.2 \times 10^{-4}$ ml $O_2$ per ml cell per second (Fenchel & Finlay, 1983), we find that the maximum radius of a spherical protozoan is about 0.8 mm, a figure fairly consistent with the size of the largest amoebae. Clearly this consideration does not exclude larger protozoa with a reduced specific respiratory rate. However, such organisms would probably be poor competitors relative to similarly sized metazoa with a vasculatory system which would allow for a higher rate of respiration and thus a higher rate of growth and activity.

For a swimming protozoan the constraints on size are more restrictive. This is because the fastest mechanism of swimming in unicellular organisms depends on cilia and they can propel organisms through water with a maximum velocity of about 0.1 cm/sec (see Section 2.2). In order to swim, it is necessary to reach a velocity which at least exceeds that of the sinking velocity. Again, for a spherical cell, the sinking velocity, v, (according to Stokes' law) is given by: $v = (2r^2(\rho\text{-}\rho_o)g)/9\eta$, where r is radius, $\rho$ and $\rho_o$ are the specific gravity of the cell and of water, respectively, g is the constant of gravity, and $\eta$ is the viscosity of water; thus, sinking velocity increases with the square of the radius. Letting $v = 0.1$ cm/sec, $\rho = 1.1$, and $\eta = 0.01$ poise, we find that the diameter of a spherical, swimming ciliate cannot exceed about 200 $\mu$m. This again is a result which is fairly consistent with nature.

These arguments, although not very precise, do show that the size range of protozoa is limited by their basic organization in conjunction with fundamental physical constraints. It is also probable that these limits have been reached by real protozoa.

# 2

# Ecological Physiology: Motility

## 2.1 Introduction

This chapter concentrates on the functional biology of protozoa, that is, the study of physiological and structural properties from the viewpoint of their adaptive significance. In protozoology the subject is intimately related to cell biology in general, as regards structure, function and biochemistry. I have tried to limit the subject to its ecological and adaptive aspects: motility, for example, will be treated with emphasis on hydrodynamics and its adaptive significance, with only passing reference to the molecular basis.

One theme which recurs in most sections of this chapter is the problem of size and scaling. Many properties of protozoa (and other microorganisms) are anti-intuitive simply because these organisms are small. Problems of scaling, especially the fact that basic functional properties do not simply scale as a function body length, have been recognized for a long time but they have only recently received considerable attention (see e.g., McMahon & Bonner, 1983, and Schmidt-Nielsen, 1984). In spite of this, false conclusions, based on erroneous assumptions about functional aspects of microorganisms, abound in the literature.

This and the following chapters will consider the motility of protozoa and the reasons why protozoa are motile. We will also look at feeding mechanisms and their adaptations to various food items, ecological bioenergetics, and the adaptive nature of different protozoan life cycles with respect to environmental patchiness in time and space. Understanding these features is of paramount importance for understanding the role of protozoa in nature. Chapter 3 will also be concerned with

the ability of protozoa to adapt to the physico-chemical environment
(including "extreme" environments), to symbiosis, and to the adaptive
nature of sexuality.

## 2.2    How protozoa move: life in syrup

All protozoa show some motility; practically all forms move freely in
the environment during some part of their life cycle, but even forms
which are usually sedentary show motility in the form of contraction
or the ability to generate water currents from which food particles may
be strained. All protozoa, of course, are motile during phagocytosis,
during the process of "cyclosis" (the intracellular movement of food
vacuoles and other organelles), and during cell division. These last aspects of motility will receive little attention: Instead we will concentrate
on swimming, the generation of feeding currents, and creeping motion
along solid surfaces.

Swimming in protozoa is often illustrated by macroscopic analogies.
In the popular book by Jahn et al. (1979), for example, ciliary swimming
is compared to human swimming, the effective stroke of a cilium being
likened to the power stroke of the swimmer. Charming as such comparisons may be, they are, in a sense, false analogies. The swimmer
moves because when he or she forces some water in one direction the
preservation of momentum will cause the swimmer to move in the opposite direction. While moving the arms forward again, coasting occurs
in the direction of swimming. This is not the case for a swimming ciliate:
It is propelled forward by the viscous (frictional) forces acting on the
cilia; inertial forces of momentum play practically no role. It can be
calculated (see Berg, 1983) that a spherical ciliate (radius: 50 $\mu$m)
swimming at a speed of 1 mm/sec would, if the cilia suddenly stopped
beating, come to a halt within $5 \times 10^{-5}$ sec, allowing it to coast for
about $5 \times 10^{-2}$ $\mu$m. Thus, coasting is not a significant factor in ciliary
swimming.

Microorganisms live in a viscous world and they are always surrounded
by a sticky coat of the surrounding water. Physicists express this by
saying that these organisms live at low "Reynolds numbers." Reynolds
number (Re) is a dimensionless quantity which expresses the ratio between inertial and viscous forces. It is equal to $(l \cdot p \cdot v)/\eta$, where l, p,
v, and $\eta$ are "a characteristic length," specific density, velocity, and
viscosity, respectively. The viscosity of water (20°C) is about 0.01 dyn
cm$^{-2}$ sec (poise). If Re $<$ 1, viscous forces predominate; if Re $>$ 1,
inertial forces are most important. For a swimming bacterium (l = 1
$\mu$m, v = 30 $\mu$m/sec), Re = $3 \times 10^{-5}$, and for a swimming *Paramecium*
(l = 150 $\mu$m, v = 500 $\mu$m/sec), Re = $7.5 \times 10^{-2}$. For a swimming

person, Re is about $10^6$. In order to make a realistic scale model of a swimming *Paramecium*, for example, in which hydrodynamic properties can be studied, its Re must be that of the real ciliate. If a swimming human is to be used as a realistic model of a *Paramecium*, the person must swim in a substrate like glycerol (viscosity: about 15 poise) in which movement will occur at about 4 mm per minute. These considerations will play a role in several contexts in this chapter. For a general (and very entertaining) discussion on "life at low Reynolds number," see Purcell (1977).

Flagella and cilia constitute the most important swimming organelles in protozoa. Both structurally and with regard to their function at the molecular level they are similar. The "sliding-microtubule model" for ciliary motion suggested by Satir in the 1960's is now generally accepted (see Satir, 1984, and Sleigh, 1974). This principle is based on the relative sliding motion of microtubules mediated by molecules of the protein dynein in the presence of ATP and magnesium ions. Hydrodynamically, however, flagella and cilia differ in function. There are usually only one or two flagella per cell whereas a ciliated cell has a large number of cilia. The motion of most flagella is characterized by undulatory waves (mostly in one plane) beginning from the base of the flagellum. A cilium has only one bend at any one time: Its movement is characterized by an effective stroke in which the cilium bends at the base while the remainder of the cilium is rather straight, and a recovery stroke, during which the cilium is drawn back to its initial position closer to the cell surface and in a plane somewhat displaced relative to that of the effective stroke. Typical beat frequencies of cilia and flagella are around 50 Hz, but much lower frequencies can be observed, especially in some flagellates. In all cases, the motility of cilia and flagella serves to propel the organism through the water or to generate water currents from which food particles may be strained.

The flagella of flagellates come in two versions—smooth ones and "hairy" or "hispid" ones. The latter type has rows of flagellar hairs, "mastigonemes," which alter their hydrodynamic properties. Flagella (and cilia) basically are able to cause movement of the protozoan because the drag on a cylinder drawn through a viscous fluid differs according to whether it is oriented parallel to the direction of movement or perpendicular to it. In the case of a smooth cylinder, the drag perpendicular to the direction of motion is about twice that of the parallel drag. In a smooth flagellum, therefore, the water will be propelled in the same direction as the waves propagating along the flagellum and the thrust on the flagellum from viscous forces of the water will act in the opposite direction. Choanoflagellates are an example of flagellates with a smooth flagellum; water is propelled away from the anterior (flagellated) pole of the cell if it is attached to a substrate, while unattached cells will move through the water with the flagellum trailing after the cell (see Figure 3.4,A).

In hairy flagella, on the other hand, the drag parallel to the flagellum is larger than the perpendicular drag. Consequently, the generated water current changes direction and moves in the opposite direction to that of the flagellar waves. Flagellates with hairy flagella, if the cell is un-attached, swim with the cell trailing after the flagellum. Hairy flagella are found in many flagellate groups including chrysomonads and its relatives, helioflagellates, dinoflagellates, bodonids, and euglenids (al-though regarding the three latter groups no exact account of how the flagella work exists). In addition, many of these possess a smooth fla-gellum which may or may not contribute to the motility of the cell (Figure 8.2,B).

In many flagellate types (such as in choanoflagellates or chrysomon-ads) in which the flagellar beat lies within one plane, hydrodynamic models can account for the generation of water currents and the velocity of the swimming cells (Holwill, 1974; Lighthill, 1976). In some cases the geometry is rather complex and a full description of the motility has not yet been obtained. In dinoflagellates there are two flagella; one is hairy and is situated in a groove along the equator of the cell, while the other is a smooth, trailing flagellum. The motility of the former leads to a rotation of the cell in the same direction as the metachronal wave passing along it (Gaines & Taylor, 1985), but the way this relates to the overall motion of cells with an asymmetrical morphology is not fully understood.

Ciliated cells are usually densely covered with cilia arranged in rows. In some ciliates these rows are arranged in regular meridians, but in many forms they are twisted in some way or parts of the cell may not be ciliated (see Figure 10.2). One of the most obvious features of a swimming ciliate is the metachronal wave generated by ciliary motion. These waves occur in different patterns according to species and swim-ming mode (see Figure 9.5,A). The explanation of this phenomenon has been the subject of lively discussion throughout the century and various neural or other control mechanisms have been suggested. It is now clear that the phenomenon is purely hydrodynamic. At the low Reynolds num-ber of an individual cilium, each one is surrounded by a coat of water which follows the movement of the cilium. When cilia are sufficiently close together, neighboring cilia will share these zones of attached water, so that their beat cycles become partly synchronized. The metachronal patterns which arise are therefore solely a function of the geometry of the surface, the position of the cilia, their beat frequency, and, of course, the viscosity of the medium, but it does not imply any sort of "nervous" control on the part of the cell (Sleigh, 1984; for a general description of metachronal patterns in ciliates, see Machemer, 1974).

Hydrodynamic models of the motility of ciliates are less complete for flagellar motility. Attempts to model the motion of the cells have been made by describing the action of individual cilia, but also by considering the actions of the metachronal waves, that is, by considering the tips of the cilia as constituting a smooth, undulating surface ("envelope models"). Both have had some success in predicting swimming speeds, but cannot yet account for the great variety in geometric design of ciliates (see Holwill, 1974, and Roberts, 1981).

A special type of ciliary propulsion is provided by the ciliary membranelles of some ciliates. Membranelles are rows (usually three, but from two to ten can occur) of very densely arranged cilia. These are not attached to each other, but due to their close proximity, hydrodynamic coupling forces them to beat fairly synchronously. Membranelles are situated in zones which consist of rows of parallel membranelles. Membranelle zones are associated with the left side of the mouth in many ciliates. In hymenostome ciliates there are usually three membranelles in the oral zone; in heterotrich, oligotrich, and hypotrich ciliates the zone contains a great number of membranelles, in some cases as many as one hundred. The main function of these oral membranelles is to propel and strain water for food particles, but in oligotrichs and in hypotrichs, for example, they also cause swimming. Some ciliates have membranelle-like structures which are not associated with the mouth and which serve only for swimming.

Although the plane of the beat of the individual cilia is perpendicular to each membranelle and thus parallel to the membranelle zone, water is propelled parallel to the membranelle and thus perpendicularly to the zone (Fenchel, 1980a). An explanation of the action of membranelle zones, based on observations on the hypotrich *Euplotes* is offered in Figure 2.1 (Fenchel, 1986a). In this ciliate, the membranelles beat with a frequency of about 50 Hz and each membranelle is about one-seventh out-of-phase relative to its neighbor so that a metachronal wave moves forward along the zone away from the cytosome, the area at which food particles are phagocytized. Within each membranelle, however, the individual cilia are also slightly out-of-phase, the innermost ones being one-seventh of a beat cycle ahead of the peripheral ones. The result of this is that metachronal waves continuously move from the peristomal cavity outward, driving water out of the cavity rather like a number of parallel, peristaltic pumps. Since each membranelle is about 10 $\mu$m wide, it is easy to calculate that the metachronal waves move along them with a velocity of about 3.5 mm/sec which is fairly consistent with the fact that the water velocity through the membranelle zone reaches about 1 mm/sec.

What sort of velocities can be achieved by organisms propelled by flagella or cilia? Empirically, it appears that the swimming velocities of

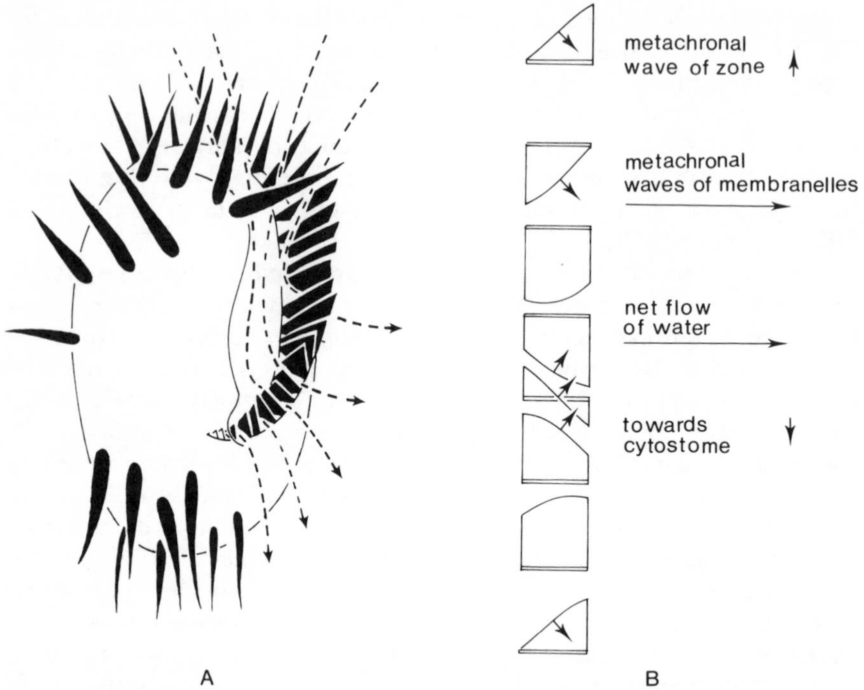

**Figure 2.1**  A. Water currents generated by the membranelles of *Euplotes*. B. The mechanism for the generation of water currents. Double lines indicate the basis of the membranelles, single lines their ciliary tips. Individual cilia move in a direction parallel to the membranelle zone, but they are slightly out of phase so that waves move along each membranelle from the peristome side and outward, thus generating water currents. (Redrawn from Fenchel, 1986a.)

both flagellates and ciliates are invariant with cell size and do not vary very much within these two types of swimming organisms. In Figure 2.2, the swimming velocity per cell length is plotted against cell length for a number of flagellates and ciliates: The data indicate that a ciliate swims with a velocity of about 1 mm/sec, irrespective of body size; for flagellates the figure is about 0.2 mm/sec. This finding would seem contrary to our intuitive observation of protozoa; since small ciliates seem to swim very rapidly, while very large ones seem to glide along slowly and majestically. However, this is because small ciliates are usually observed at higher magnifications than large ones and because one tends to evaluate speed as a function of the size of the moving object rather than according to an absolute scale.

This empirical finding of the invariance of velocity has been rationalized as follows (Sleigh & Blake, 1977): The speed of propulsion by

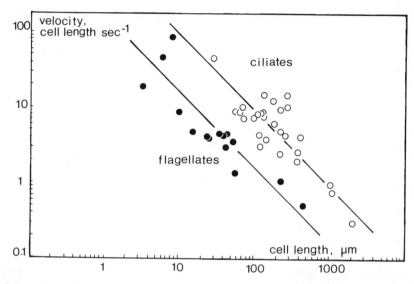

**Figure 2.2**   The swimming velocity of fifteen species of flagellates and twenty-eight species of ciliates (expressed as cell lengths per second) plotted against cell length. Slopes of −1 indicate constant, absolute swimming velocities independent on cell length. (Altered from Sleigh & Blake, 1977.)

a single cilium must be proportional to its angular velocity and length. The force is then proportional to the angular velocity, the square of length, and viscosity. Finally, the necessary power required is proportional to the square of the angular velocity, the third power of length and viscosity. The power therefore rapidly increases with angular velocity and length and this might limit the increase in propulsion velocity. There are also structural constraints on the length of a cilium (or flagellum). The observed minimum radius of curvature is around 2 μm, and this would limit the minimum length of a motile cilium to about 5 μm, which is a little less than what is actually found in some small ciliates. Similar constraints probably apply to the maximum length of cilia and flagella, which rarely exceed about 50 μm. In typical flagellates the one or few flagella are much longer than the cell; in ciliated forms the converse is true; also flagellates are generally much smaller organisms than ciliates. Some very large flagellates, such as hypermastigines living in the intestine of termites and the opalinids living in the intestine of amphibians, are in fact covered by numerous flagella or cilia. These constraints on swimming velocities are, of course, of significance for understanding the characteristics of protozoa with regard to dispersal and food particle concentration in nature.

There are a number of less-common swimming mechanisms in protozoa which do not rely on flagella or cilia. In part these can be understood as mechanisms for escaping the limitations of ciliary motility in large cells. Some dinoflagellates (e.g., *Erythropsis*, see Figure 1.3) propel themselves with a tentacle which contracts periodically, acting rather like a piston. Other protozoa (e.g., the dinoflagellate *Noctiluca*) propel themselves by deformations of the entire cell brought about by contractions of "myonemes"—contractile fibrilla. In the peculiar oceanic heliozoan *Sticholonche*, the spicules are moved like oars to produce swimming (see Febvre-Chevalier & Febvre, 1982, and references therein). A species of *Vorticella* uses the contractile stalk for swimming rather than for permanent attachment like other vorticellids and the pelagic oligotrich *Tontonia*, has a posterior, long appendage for swimming.

Many protozoa creep or slide along solid surfaces rather than moving by swimming. This also applies to some organisms which use flagella or cilia for locomotion. In some bodonid and euglenid flagellates a smooth, "trailing" flagellum slides along in very intimate contact with the substratum, while the hairy, anterior flagellum pulls the cell along in a manner similar to what is seen in their freely swimming relatives. These sliding flagellates can also free themselves from the substratum and swim away. The forces which hold the trailing flagellum to surfaces are not known, but it is reasonable to assume that van der Waals forces play a role. Many ciliates slide (e.g., cyrtophorids and loxodids) or walk (hypotrichs) on solid surfaces using cilia or, in the latter case, "cirri" (which are dense bundles of cilia). In addition, some forms can also attach temporarily to a solid surface using "thigmotactic" cilia. This latter term does not constitute an explanation, but again the phenomenon probably involves van der Waals forces.

The sarcodines as well as a few organisms classified as flagellates use pseudopodia for locomotion. These are temporary extensions of the cells. The motility mechanism is now believed to involve actin filaments which can slide relative to one another, mediated by myosin molecules and in the presence of calcium ions and ATP. The generated force may act on the cell membrane or on microtubules. The mechanics of pseudopodia formation, however, is still not understood in detail (Capuccinelli, 1980). There are several different types of pseudopodia: Lobopodia, characteristic of the amoebae, are thick structures and there are usually only one or a few per cell at any one time (Figure 2.3,A,B). Filopodia are very thin and numerous; they occur in organisms like *Euglypha*. Reticulopodia are anastomosing networks of very slender pseudopodia supported by microtubules; they occur, for example, in the foraminifera (see Figure 3.10,B,C). Finally, axopodia are straight and somewhat stiff structures supported by bundles of microtubules. They occur in heliozoa, radiolaria, and acantharia (Figure 2.3,C-F). All types of pseudopodia also play a role in catching food particles and

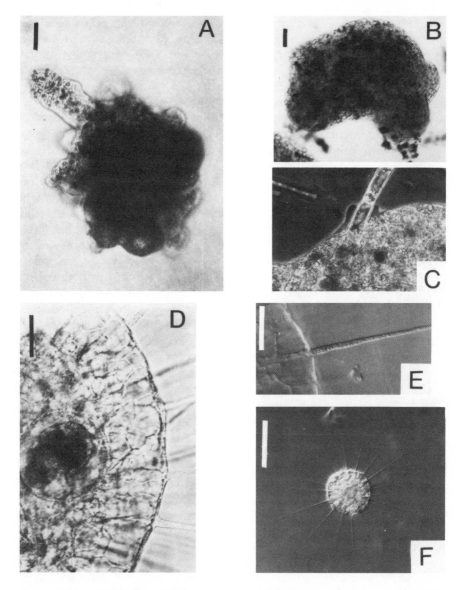

**Figure 2.3**  A. *Amoeba proteus*, scale bar: 100 μm. B and C. *Pelomyxa palustris*, in C ingesting an algal filament. Scale bar (in B): 100 μm. D and E. The heliozoan *Actinosphaerium eichorni*, showing the actinopods and (in D) an ingested rotifer; scale bars: 50 μm. F. The heliozoan *Actinophrys sol*; scale bar: 50 μm. (C: photograph by B. J. Finlay.)

axopodia are not always involved in locomotion. Some heliozoa roll along the substratum by alternately shortening and extending their ax-opods. This is the result of the assembly and disassembly of microtu-bules. In other types of pseudopodia, locomotion takes place by the extension of a pseudopod, attachment to the substratum, and eventually a contraction of the pseudopod. The details of the process, however, are often poorly understood. Amoeboid motility is much slower than that of ciliary swimming; velocities of 5 to 20 $\mu$m/sec are typical.

Actomyosin microfilaments and, in some cases, pseudopodia-like structures are also involved in phagocytosis and pinocytosis and this motile principle is also responsible for the movements of organelles within the cells. Contraction requires calcium ions and ATP. Myonemes are bundles of filaments which allow the contraction of the cells; they are typical of heterotrich ciliates, euglenid flagellates, and some other forms. In most cases they also seem to be based on actomyosin micro-filaments. In other cases, however, motile principles which are based neither on microtubule sliding nor actomyosin principles occur. One example is the contractile stalk of sessile peritrichs. In the stalk, the contractile filament, the spasmoneme, consists of a protein called spas-min, which coils up in the presence of $Ca^{2+}$, but the coiling process does not require ATP (Amos, 1975). For general discussions of the cel-lular and biochemical basis of motility and for references, see Cappuc-cinelli (1980).

One important aspect of motility is, of course, its control. The con-trolling factor seems in all cases to be the internal concentration of $Ca^{2+}$ and fluxes of this divalent ion are related to changes in the cell mem-brane potential. In motile systems based on microtubule sliding, such as cilia and flagella, increased levels of $Ca^{2+}$ slow down or reverse the beat cycle, whereas hyperpolarization (and the associated efflux of $Ca^{2+}$) leads to an increase in beat frequency. In actomyosin microfilament-based motile systems, the influx of $Ca^{2+}$ leads to contraction and the presence of the ion in the medium is necessary for the induction of food vacuole formation and for the production of pseudopodia (Cap-puccinelli, 1980; Naitoh & Eckert, 1974; Naitoh & Sugino, 1984).

## 2.3    Orientation in the environment

There are apparently only two reasons for protozoan motility: to catch food particles and to move to a new location. We will first consider the latter aspect of motility. The ability of protozoa to leave adverse envi-ronments and to congregate in more attractive ones is well documented. A common classroom demonstration is to inject a tiny drop of acetic acid into a somewhat larger drop of a suspension of paramecia. After a few minutes, the cells will congregate in a ring around the drop of acid at a location corresponding to their preferred environmental pH value. Similar effects can be seen in oxygen gradients forming around air bubbles in microscopic preparations. Photosynthetic protists (and protozoa which contain symbiotic, photosynthetic cells) will accumulate in the light.

Such phenomena are, often incorrectly, referred to as "taxic behavior" (e.g., chemotaxis, phototaxis, etc.) and sometimes the term is even used in place of an explanation of what mechanism the organism actually uses to orient itself in its environment. It will be useful to start the discussion by considering some possible mechanisms. Following Fraenkel & Gunn (1940) and Lapidus & Levandowsky (1981) I consider only oriented responses to be "taxes": the organism "knows" the direction of a stimulus or of a gradient of some stimulus. In established cases, taxic behavior is associated with some sort of complex sensory organelle which can analyze the stimulus. For example, in the case of chemical gradients, a taxic response requires that the cell be able to detect a difference in the concentration of a substance between the front and the rear end of the cell. Given very steep chemical gradients, such a mechanism may be possible for very large sarcodines or for those with very extensive pseudopodia, but, in general, it is probably safe to state that chemotaxis does not occur in protozoa (Lapidus & Levandowsky, 1981). Taxic behavior in general is not very common in protozoa. (Microbiologists use the term chemotaxis for all responses to chemical gradients; in fact, the only established case of taxic response in a prokaryote is probably the magnetotaxic behavior described in some microaerophilic bacteria; see Blakemore et al., 1980).

However, other options are available to protozoa for orientating themselves in the environment. The motility of protozoa (and other unicellular organisms) can be described as a "random walk" (see Berg, 1983). If a single cell, such as a ciliate, is followed over a period of time one observes that more or less straight "runs" are interrupted by "tumbles," in which the cell stops for a short moment and then continues in another direction. The new direction may be random or it may be related to the previous swimming direction. Tumbles are brought about by spontaneous depolarizations of the cell membrane, which leads to an influx of calcium ions and a ciliary reversal. After the membrane potential is restored, the cell resumes forward swimming. If a whole population is observed over a time scale which is reasonably long relative to the expected time interval between tumbles, and in a container which is reasonably large relative to the average run length, then the motility may be described as a diffusion process and quantified by means of a diffusion coefficient, D. If the cells at time zero are concentrated at one point in a capillary tube, their distribution at time t will be normal and with a variance of Dt. The diffusion coefficient is proportional to the square of the swimming velocity and to the average time interval between tumbles.

Protozoa can modify their motility (diffusion coefficient) according to the environmental conditions by changing the parameters of motility: swimming speed and the frequency of tumbling. This type of response is called "kinesis." One may distinguish between "orthokinesis" in which speed is modified and "klinokinesis" in which the frequency of tumbling is modified, but in protozoa, the increase in swimming speed correlates with the expected time interval between tumbles so that the two are related. (In bacteria, it seems, only klinokinesis occurs; the speed of swimming cannot be modulated.)

Cells will tend to aggregate wherever the motility (D) has the lowest value; in fact it can be shown that the steady-state distribution of the cells will be inversely proportional to their local motility (Okubo, 1980). In a container with length, l, and with two equally large environmental patches characterized by cell motility values of $D_1$ and $D_2$, the fraction of cells in patch 1 will be $D_2/(D_1 + D_2)$ when a steady-state distribution has been reached. The time taken to reach this steady-state distribution will be of the order of $l^2/D$.

Kinetic responses may be permanent; that is, under certain adverse conditions a constant, high motility is maintained by the cells, but in some cases, an "adaptation" to the environment takes place so that the cells eventually change their motility parameters back to some normal state, even if the environment does not change.

The third category of behavioral response to consider is what I will call a "transient response" (see Fenchel & Finlay, 1986b). One type of such transient response is well documented in the literature, namely, the "phobic response" or "avoidance reaction." An avoidance reaction of ciliates can be observed when a cell suddenly enters an area of adverse conditions. This immediately leads to cell membrane depolarization and ciliary reversal (Naitoh & Eckert, 1974). This behavior increases the probability that the cell will move back to the more attractive zone. The response is transient (on the order of a few seconds); if it were not, it would be dysgenic. This is due to the fact that if the cell does not escape the adverse environment immediately, it will risk being trapped there if it maintains a high frequency of tumbling (see the discussion on kinesis above). The opposite of a phobic response also occurs. If a cell enters a more attractive zone, tumbling is suppressed, and the swimming velocity increases temporarily; this decreases the probability that the cell will return to the less attractive zone.

Such transient responses may guide cells along environmental gradients if these are sufficiently steep to induce the response within one length of a run. If, for example, a cell is situated in a gradient of an attractant, and if swimming up the gradient suppresses tumbling and swimming down the gradient induces tumbling, then the diffusion coefficient will be different in the two directions and the cells will tend to move up the gradient. This type of response, which is well documented (see below), may resemble a taxic response, but it does differ in that it—as is the case for kinetic responses—is a statistical phenomenon and that the cell does not at any moment "know where it is going." Transient responses imply a sort of "memory," that is, a comparison between environmental conditions within a short time interval, roughly corresponding to about the length of one run.

One way to visualize the difference between taxic responses and the two other types is to take a glass container with a suspension of green flagellates. On one side of the container a beam of light shines through a convex lens so that the light is focused at a point in the middle of the tank. Flagellates which employ phototaxis will then gather at the wall of the container closest to the lens, although by doing so they congregate where the light is least intense. This is because they orient themselves relative to the direction of the light. Conversely, the flagellates which use kinetic and transient responses will accumulate in the middle of the container where the light is most intense.

Together these kinetic and transient responses allow for a seemingly rather complex behavior. Nevertheless, in all cases the transduction of sensory signals probably leads to changes of the cell membrane potential, which again modifies ciliary or flagellar beat patterns or the formation and contraction of pseudopodia.

In order to summarize and compare the basic properties of the three types of responses, the following can be said:

1.  Taxic responses are more efficient over larger distances since the time taken to reach the desired place is only proportional to distance and inversely proportional to swimming speed (or rather to some fraction of the swimming speed since protozoan motility always has a certain random element). The relatively rare occurrence of the taxic response is due to fundamental structural and functional constraints, especially the small size of protozoa which does not allow the sensing of chemical gradients directly and the low Reynolds numbers which exclude the "rheophilic" responses (e.g., moving upstream or against the wind) so common in metazoan animals.

2.  Kinetic responses are fast over small distances, but their efficiency decreases with the square of distance. They are the only means of orientation among large environmental patches, each of which is homogeneous with respect to environmental conditions. Over very small distances (measured relative to the average run length)

and in steep gradients, kinetic responses represent a rather inaccurate type of orientation.

3.  Transient responses, finally, represent the most accurate orien-
    tation in small environmental patches and in steep gradients.
    (Transient responses are not easy to analyze mathematically, but
    they should be suited for computer simulations. This has so far
    not been done and, unfortunately, most experimental data dem-
    onstrating this type of behavior is of a rather qualitative nature.)

Chemosensory behavior has been documented for a large number of
protozoan species. It appears that receptor molecules are situated in
the cell membrane and the motor response is always associated with
bioelectrical membrane phenomena. The nature of the receptor molecules are not, in general, known in detail. In the case of oxygen sen-
sitivity in *Euglena* and *Loxodes*, it has been suggested that cytochrome
oxidase is the receptor molecule since the response is inhibited by
cyanide (Miller & Diehn, 1978; Finlay et al., 1986).

*Paramecium* is the classical organism for behavioral studies in protozoa and its chemosensory responses are well documented. Kinetic as
well as transient responses occur. Most experiments have been made
with attractant and repellent test solutions, such as K-acetate and BaCl$_2$;
the responses are not easy to interpret from an ecological point of view.
It has been shown, however, that this bacterivorous ciliate is attracted
to bacterial extracts (Van Houten et al., 1981; Antipa et al., 1983).
*Tetrahymena*, which in nature feeds on patches of decaying, organic
material (some species can also attack invertebrates which they then
devour), is attracted to a variety of amino acids (Almagor et al., 1981;
Levandowsky et al., 1984). In this case also, kinetic as well as transient
responses guide the cells toward the source of the attractant. This sort
of chemosensory behavior is obviously adaptive for the numerous species of protozoa which depend on patchy food resources.

Chemosensory behavior is also important in sexual processes. The
preconjugants of many species of ciliates and the gametes of some flagellates have been shown to excrete water-soluble "gamones." These
gamones are mating-type specific and attract other mating types of the
same species (Miyake, 1981; Van Houten et al., 1981).

Many protozoa respond to oxygen, which may be either an attractant
or a repellent, according to species and to ambient oxygen tension. The
response to oxygen and the ability of many species to avoid anoxic
conditions is readily observed in containers with decaying material.
Here the protozoa will usually congregate at the top if this has direct
access to the air. Many sediment-dwelling protozoa show vertical zonation patterns (see Figure 8.8), which at least in one case, namely,
loxodid ciliates, is due to a direct response to oxygen. If the sediment

becomes totally anoxic these ciliates migrate up into the water column. *Loxodes* is a microaerophilic species which prefers oxygen tensions around 5 percent atmospheric pressure. The response to oxygen is complex. At $O_2$ levels exceeding the optimum, the cells become positively geotactic (see below) and they also increase swimming speed and suppress tumbling. A sudden increase in the $O_2$ level induces an avoidance reaction. In anoxic water the ciliates display negative geotaxis and also increase swimming speed relative to that of optimum conditions. These behavioral responses can together account for the distribution patterns of the organism in oxygen gradients. The fact that these ciliates avoid high oxygen levels can be partly explained by the intracellular production of toxic oxygen radicals. Their production is apparently enhanced photochemically in these pigmented ciliates, and in the presence of even small amounts of $O_2$ they respond to light with a negative geotaxis as if the oxygen concentration had been increased. This negative response to light is inhibited by cyanide and the light receptor is therefore likely to be identical to the $O_2$ receptor (Fenchel & Finlay, 1984, 1986a,b; Finlay, 1981; Finlay, Fenchel & Gardner, 1986; see also Figure 9.7).

*Paramecium bursaria* normally contains symbiotic *Chlorella* cells. These ciliates are attracted to light, but they lose this behavior if they do not harbor symbionts. Their light responses include both kinetic and transient ones, and they are mediated, at least in part, by oxygen produced by the symbionts in light (Cronkite & Van den Brink, 1981).

Many other protozoa also respond to light. The phenomenon has been studied best in pigmented ciliates. Within the heterotrichs, but also in other groups, brightly colored species occur, which owe their color to pigment granules situated beneath the cell membrane. The pigments are either flavins, hypericins, or both, and their adaptive significance is unknown. Their presence correlates with strong negative responses to light. Detailed studies have been made on *Blepharisma* and *Stentor* (Matsuoka, 1983; Pill-Soon & Walker, 1981). These ciliates display strong phobic reactions when the light intensity is suddenly increased and at least *Blepharisma* also displays a kinetic response (increased swimming speed) if the strong light intensity is sustained. More recently, it has been shown that the unpigmented *Paramecium multimicronucleatum* shows similar responses (Iwatsuki & Naitoh, 1983). Whether these examples have a similar basis (viz., oxygen toxicity), as in *Loxodes*, is not known.

Photosynthetic protozoa are attracted to light. Pigmented flagellates usually carry an "eye spot," a cluster of carotenoid droplets situated close to the anterior end of the cells. This, however, is not believed to be the photoreceptor, which is situated at the base of a flagellum. In *Euglena* it has been suggested that locomotor responses are triggered when the eye spot casts a shadow on the base of the flagellum. In this way the photoreceptor and the eye spot together analyze the direction of the light rays, explaining the true phototaxic response in this flagellate. For discussion of the photoresponses of *Euglena* and of other photosynthetic protists, see Foster & Smyth (1980) and Pill-Soon & Walker (1981).

Mechanoreception is common in protozoa and ciliates in particular often show ciliary reversals if they collide with solid objects. Cilia have long been thought to play the role of mechanoreceptors, especially the immobile cilia, such as the short, bristle-like forms on the dorsal side of *Euplotes* (Figure 1.2). However, it is now known that the cilia only act as levers and that the transduction of the mechanical signal is actually due to mechanical deformation of the cell membrane, which leads to a change in membrane potential and thus to a locomotor response. In the ciliates *Dileptus* and *Paramecium* it has been demonstrated that the electrical response of the membrane upon mechanical deformation differs between different regions of the cell. If a ciliate is mechanically stimulated at its posterior end, a hyperpolarization, and thus an increased swimming speed follows, whereas if a cell is stimulated at its anterior end, a depolarization, and hence a ciliary reversal follows (Naitoh & Eckert, 1974).

Geotaxic behavior in protozoa has been a much debated subject for a large part of the century (see Roberts, 1981, for references). It is true that some ciliates tend to swim upwards in containers, but the organisms may, under certain environmental conditions, reverse their "geotaxic" response and congregate at the bottom of the container. It is, however, difficult to explain these observations since protozoa have no organelle which can be interpreted as a gravity sensory, although it has been suggested that vacuole contents could act as a statolith, a dense body, the gravitation induced movement of which is sensed. It can be shown that a gravity sensor must have certain minimum dimensions, because otherwise, Brownian motion will override any effects of gravity (Fenchel & Finlay, 1984).

Organelles involved in geotaxis should therefore be easy to observe. In the case of the apparent geotaxis of *Paramecium*, Roberts (1970) gave an ingenious explanation: This ciliate does, like most other protozoa, have a specific gravity slightly above unity so the cells will slowly sink due to gravity. However, due to the characteristic shape of the cell, the posterior end being more bulky than the anterior one, they tend to orient themselves so that the anterior end eventually points upward as they sink downward. This can be observed in anesthetized cells and in scale models immersed in a viscous fluid; it is a purely hydrodynamic phenomenon. If the ciliates tumble sufficiently rarely, the cells will tend to swim with the anterior end pointing upward and will eventually

congregate at the top of the container. However, if the tumbling frequency and therefore the rate of random reorientation increases, or if
swimming speed decreases, the effect of gravity overrides the tendency
to orient the anterior end upward and the cells will go to the bottom.
The paramecia can therefore, to some extent, control whether they move
upward or downward in the water column, but it is not a true geotaxis
in the sense that they can detect the direction of gravity.

It remains to be seen how general Roberts' model is. There are many
examples of protozoa which seem to be capable of controlling the level
at which they are found in the water column, but in many cases this
may be due to light responses or to orientation in chemical gradients.
The only experimentally established case of a true geotaxis is that of
*Loxodes*. In this case, a complex mechanoreceptor is present which can
account for the behavior (Fenchel & Finlay, 1984, 1986a; see also Figure
1.3), but similar organelles have not yet been found in other protozoa.

# 3

# Ecological Physiology: Feeding

## 3.1 General considerations

Phagocytosis is an essential feature of protozoan feeding and it is the subject of both this and the following section. Some protozoa acquire additional energy and materials for growth by other mechanisms, such as the uptake of dissolved materials, which will be discussed at the end of this section, while symbiotic relations with photosynthetic organisms will be discussed in Section 6.2.

Feeding consists of two processes, each of which may limit the actual rate of feeding. The first one of these is the process of phagocytosis, that is, the enclosure of a food particle in a membrane-covered vacuole in which digestion takes place. In ciliates and in most phagotrophic flagellates this occurs at a special site on the cell surface, the "cytostome," which is covered by a single unit membrane from which food vacuoles are formed. The cytostome is often associated with various organelles in the surrounding cytoplasm, in particular bundles of microtubles, which play a role in the transport of the captured particles. On the surface, surrounding the cytostome, a variety of ciliary and other organelles serve to concentrate or retain food particles. The entire area is usually referred to as the "mouth." For ciliates, a rather specialized terminology of mouth morphology has developed due to its taxonomic significance (see Corliss, 1979). However, in sarcodines, phagocytosis seems to take place anywhere on the cell surface.

The process of phagocytosis has been extensively studied in ciliates (Allen, 1984; Nilsson, 1979, and references therein) but only more rarely in other types of protozoa (e.g., Linnenbach et al., 1983). The

32

induction of phagocytosis is, at least in part, due to mechanical stimulus by food particles. Food vacuole formation does not take place in particle-free water, while filter-feeding protozoa ingest inert particles such as latex beads and suspended carbon particles at a rate similar to that of similarly sized food particles (Fenchel, 1986a; Mueller et al., 1965). The newly formed food vacuole may be subject to some water absorption, after which it fuses with primary lysosomes to form secondary lysosomes in which hydrolytic degradation of the food particles takes place. The dissolved nutrients are removed from the vacuole through pinocytotic vesicles. Indigestible remains are eventually removed from the cell through a fusion of the vacuole with the surface membrane; in ciliates this takes place at a special site, the "cytoproct." The last part of this cycle is the transport of membrane material in the form of small vesicles back to the cytostome area to form new vacuoles. Thus, during the lifetime of a food vacuole, it is moved around in the cell. Individual food vacuoles are easy to follow when labelled with inert, light-refringent or colored particles. It seems, at least in ciliates, that the elimination of food vacuoles is random, so that the lifetime of vacuoles has an exponential distribution. The expected lifetime of a vacuole decreases with increasing rate of feeding, so that if the cells are starved the elimination rate is much lower. The average lifetime for a food vacuole in a feeding ciliate is about twenty minutes (Berger & Pollock, 1981).

The maximum rate at which phagocytosis can take place (which may be limited by the rate at which membrane material can be recycled within the cell) sets an upper limit to the feeding rate of a protozoan. The maximum volume ingested by a small ciliate or flagellate is about 100 percent of the cell volume per hour, which is consistent with minimum doubling times of such organisms of about three hours. In larger protozoa this figure is lower; about 50 percent of the body volume per hour (Fenchel, 1980c, 1982b).

The other process involved in feeding is that of concentrating food particles from the environment. It is possible that the first Precambrian protozoan simply engulfed portions of the surrounding water and ingested whatever bacteria were present, but extant forms have developed various mechanisms by which the dilute food particles of the environment can be concentrated prior to phagocytosis. The variety of these adaptations contributes to the diversity of protozoan forms; this will be discussed in the following section, while we devote this section to some general aspects of particle capture.

The first thing to consider is how to quantify the rate at which the organisms concentrate particles. A reasonable measure is "clearance," F, by which is meant the volume of water cleared of food particles per unit time. In discussing clearance, we will, for the moment, consider

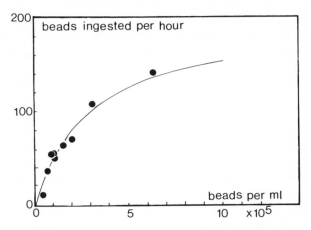

**Figure 3.1**    The uptake of 2-μm-size latex beads by the oligotrich ciliate *Halteria grandinella*, as a function of particle concentration. The data are fitted to a hyperbolic function. The slope at the origin is the rate of maximum clearance which is $6.7 \times 10^{-4}$ ml/h. The volume of the ciliate is about $8 \times 10^3$ μm³, so it clears about $8 \times 10^4$ times its own cell volume per hour. The maximum uptake rate is about 190 beads per hour. (Data from Fenchel, 1986a.)

only organisms which feed on suspended particles, although the principles discussed below can easily be adapted to organisms which feed on particles associated with solid surfaces. Clearance is then equal to the particle uptake per unit time, U, divided by particle concentration in the surrounding medium, x, or: $F = U(x)/x$. The rate of food uptake, however, is not likely to increase linearly with particle concentration, since as x increases, the rate of phagocytosis becomes limiting. Assuming that it takes a finite time, $t'$, to phagocytize one unit of food particles, during which additional phagocytosis cannot take place, the rate of food uptake as a function of food particle concentration (the "functional response") then becomes: $U(x) = xF_m (1 - t'U)$, where $F_m$ is a maximum value of clearance realized for very low values of x. Rearranging the equation we get: $U = xU_m/(x + U_m/F_m)$, where $U_m$ is the maximum rate of phagocytosis ($= 1/t'$).

This equation is a hyperbolic function in which, as x becomes very large, $U = U_m$; the slope at the origin is $F_m$ (Figure 3.1). This is analogous to the Michaelis-Menten equation for describing enzyme kinetics, which also describes, for example, the rate of uptake of a dissolved nutrient by a bacterial cell as a function of concentration. This way of deriving the equation shows that the "half-saturation constant" ($U_m/F_m$) is an ad hoc parameter which has no obvious biological significance. The half-saturation constant can be interpreted as the ratio between the capacity to ingest particles and the efficiency with which particles are concentrated from the environment. This is an analogy to the uptake of dissolved molecules by a bacterial cell. Here the half-saturation (Monod) constant measures the ratio between the transport capacity of the cell membrane and the uptake rate when only limited by the rate of diffusion outside the cell. Also in this case, it is the latter measure and not the half-saturation constant which gives information on the competitive ability at very low substrate concentrations (Koch, 1971). The significant parameters are $U_m$ and $F_m$; the latter (and not the ratio $U_m/F_m$) is a direct measure of competitive ability for scarce resources. This has been widely misunderstood in the literature and has led to incorrect conclusions (as discussed below) about the role of dissolved organic materials as food for protozoa.

The actual values of $U_m$ and $F_m$ may be determined in various ways (Fenchel, 1980b, 1986a). One is to measure the uptake of either real food particles or inert particles (e.g., latex beads) as a function of concentration. It is also possible to calculate the values from culture experiments since the exponential growth rate constant is proportional to the food consumption rate (the growth yield being the proportionality constant). Examples of such experiments are shown in Figures 3.1 and 4.1. For any one species $F_m$ and $U_m$ are functions of particle quality. In particular, particle size affects both the efficiency of retention and the rate of phagocytosis.

The "volume-specific clearance," that is, clearance divided by cell volume, is often a useful number, since it allows a comparison of clearance values between different-sized organisms. If the clearance of an organism has been determined and the particle concentration in the natural environment is known, it is possible to estimate the consumption rate in nature. Values of maximum volume-specific clearance for protozoa feeding on suspensions of particles are usually around $10^5$ per hour; that is, the protozoa can, in an hour, clear food particles from a volume of water which is $10^5$ times their own cell volume. Species which filter very small particles or those which live in interstitial environments with narrow crevices have values of clearance which are as much as ten times lower.

What mechanisms do protozoa utilize for concentrating food particles? At the dimensions and velocities characteristic of protozoa, any mechanism based on inertial forces can be ruled out; some quantitative calculations demonstrating this are given in Fenchel (1980a). For organisms feeding on suspended particles there are three possibilities; these mechanisms can be termed "filter feeding," "direct interception," and "diffusion feeding" (Fenchel, 1984, 1986a). These mechanisms are analogous to methods for catching fish, namely, trawling, spear fishing, and the use of fish traps. Filter feeding depends on the transport of water through a filter formed by cilia or pseudopodial tentacles, which strain food particles from the water. In direct interception or "raptorial feeding," food particles carried along the flow lines are directly intercepted by the protozoan surface which is "sticky," thus retaining the particle until it is phagocytized. These two mechanisms depend on the motility of the protozoan. The third mechanism, diffusion feeding, requires motility of the prey, which is intercepted by even a motionless consumer.

In order to get a feeling for the relative efficiency and importance of these mechanisms, it is useful to consider very simplified models of suspension-feeding protozoa. In the case of a filter-feeding protozoan, the capture rate is proportional to the concentration of food particles, to the area of the filter, and to the velocity of the water current which the cell generates: Thus, the capture rate is $xR^2v$; where x is the particle concentration in the environment; R, a length measure of the cell; and v, water velocity. Volume-specific clearance is then found by dividing with x and with cell volume, which is proportional to $R^3$, to give the expression $R^{-1}v$. (Any realistic model which is to predict the clearance rate of a real protozoan must, of course, take into account details of the geometry of the cell as well as hydrodynamic considerations. However, the above expression yields results of the correct order of magnitude: If $v = 300$ $\mu$m/sec and $R = 10$ $\mu$m, then volume-specific clearance is about $10^5$ per hour.)

For a spherical protozoan (radius: R) catching particles (radius: r) by the second mechanism of direct interception, there exist "critical flow lines" within which particles will be intercepted. The transectional area of the flow past the cell within the critical flow lines is $2\pi Rr$, (if R $>>$ r, and if it is assumed that particles can be intercepted along the equator of the cell), so the volume-specific clearance will be proportional to $R^{-2}rv$ (based on arguments quite similar to those applied for the filter feeder above). The efficiency of a raptorial feeder therefore depends on the size of the prey. To compare the efficiencies of a filter feeder and a raptorial feeder of equal size and capable of generating a similar water velocity, we can divide the expression for clearance of the latter by that for the former to arrive at the expression, r/R. This informs us that efficiency of removal of particles is a function of the ratio between predator and prey size. If the food particles are sufficiently small, filter feeding is the most efficient mechanism. Empirical evidence for ciliates shows that if the prey:predator size ratio exceeds about 0.1, raptorial feeding predominates, but if the food particles are smaller, filter feeding is found (Figure 3.2, Fenchel, 1986a). Thus, many smaller, predatory ciliates (e.g., *Litonotus, Didinium*) feed on other ciliates such as *Colpidium* and *Paramecium* by direct interception, while the giant ciliate *Bursaria* (which measures up to 1 mm) feeds on similar prey organisms by filter feeding.

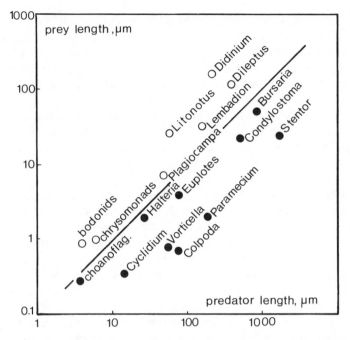

**Figure 3.2**   Average food particle size as a function of cell size for a number of filter-feeding (filled circles) and raptorial (open circles) protozoa. The line corresponds to a food particle:cell size ratio of 1:10. (Fenchel, 1986a.)

The last mechanism is feeding by diffusion. It is related to the mechanism by which a bacterium takes up organic molecules from a solution. It can be shown (Koch, 1971; Roberts, 1981), that at low particle concentrations (the diffusion-limited case in which the particle concentration at the surface of the organism is zero) the uptake rate by a spherical collector is $4\pi RDx$, where D is the motility (diffusion coefficient) of the particles (or molecules): Hence the volume-specific clearance is proportional to $R^{-2}D$.

The efficiency of the mechanism depends on the motility of the prey. For small bacterivorous heliozoa, Fenchel (1984) calculated that in the case of nonmotile bacteria, Brownian motion alone will yield a clearance which is two to three orders of magnitude too low to make these protozoa competitive with other bacterivorous forms. If motile bacteria are considered, however, the efficiency of the mechanism is comparable to that of protozoa utilizing direct interception (e.g., chrysomonad flagellates) or filter feeding (e.g., choanoflagellates).

We have so far considered only suspension-feeding forms. For forms which feed on food particles associated with solid surfaces, filter feeding does not occur. However, it is still possible to distinguish between forms which slide along the surface in order to intercept prey and those which spread their pseudopodia over a large area to trap motile prey.

There are a number of mechanisms which may enhance the capture of particles. One of these is to attract the prey (or to return to the fishing analogy, to use bait). Apparently some foraminifera and heliozoa excrete substances that attract flagellates, which then can be captured by the predators (Lee, 1980, and references therein). Unfortunately, the phenomenon has not been studied in detail. Certain physical principles could possible act to increase the probability of capture in protozoa where food particles stick to the collector surface, i.e., electrostatic attraction between food particles and the predator surface. Another such mechanism is based on the observation that particles of a finite size following a viscous flow close to a solid surface will occupy a part of a velocity gradient. This will lead to a rotation of the particle which will tend to migrate across the flow lines and toward the surface (Spielman, 1977). The significance of such effects for suspension-feeding organisms has not yet been investigated. In the case of filter-feeding protozoa, however, they can be ruled out since food particles are not trapped by mucus, extrusomes, or other mechanisms prior to phagocytosis (food particles show Brownian motion in the food vacuoles); so in this case, a purely mechanical sieving effect accounts for particle capture.

The simple models discussed above ignore certain hydrodynamic constraints which apply in particular to filter feeders. These constraints are related to the fact that when a viscous fluid flows past a solid surface, the velocity is zero at the surface (the "no-slip condition") and a steep velocity gradient is found immediately above the surface. One consequence of this (which is well known from the practical application of filters in the laboratory) is the pressure drop across a filter necessary to overcome viscous forces. Ciliary propulsion mechanisms can only sustain a very low hydrostatic pressure. The pressure drop across a filter is proportional to fluid velocity, viscosity, and the thickness of the filter and it is also a rather complex function of the geometry of the filter elements; in general, it increases greatly when the size ratio between the filter elements and the pores becomes large ($>1$). The pressure drop of some protozoan filters (those consisting of parallel arrays of cilia or tentacles) can be estimated. This pressure drop is always about 10 dyn/cm$^2$ (0.1 mm H$_2$O). Since the diameter of the filter elements is fixed at 0.2 $\mu$m (cilia) or down to 0.1 $\mu$m in some tentacles, it means that the flow rate through the filters (and thus clearance) is much reduced in filter-feeding species which specialize in feeding on very small food particles. For example, choanoflagellates, which filter the smallest prokaryotic cells (the distance between neighboring pseudopodal tentacles is only 0.1 to 0.3 $\mu$m) have flow rates through the filter of only

10 to 20 $\mu$m/sec; whereas in the helioflagellate, *Pteridomonas*, the free distance between the tentacles is 1–3 $\mu$m and the flow velocity through the filter is nearly 100 $\mu$m/sec. The specialization to catch very small food particles is therefore correlated with a decrease in clearance (Fenchel, 1986a).

The flow past a filter feeder is larger when the organism is attached to a surface than when it is swimming. This is because the thrust of the flagellum or cilia of a swimming organism must balance the viscous drag of the cell as it moves through the water. In accordance with this, most filter-feeding protozoa tend to attach to a solid object when feeding. Some forms are permanently attached (e.g., peritrich ciliates and some choanoflagellates) and others attach temporarily while feeding (e.g., many ciliates such as *Cyclidium* or *Euplotes*). This, however, poses another problem for the protozoan, namely, the viscous drag due to the proximity of a solid surface, an effect which could slow the water flow considerably (Fenchel, 1986a). It can be shown that for a given distance from the substrate, the effect is considerably larger if the feeding current is perpendicular to a solid surface than if it is parallel. Thus, choano-flagellates, *Vorticella* and *Stentor*, which propel water perpendicular to the surface, have stalks, while organisms such as *Euplotes* and *Cy-clidium*, which propel water parallel to the surface, only need cilia to rise sufficiently far above the surface. The effect of a solid surface is a function of the ratio of the distance to the surface and the radius of the filter. When, in the case of a perpendicular flow, this ratio exceeds eight, the effect of a solid surface becomes negligible, and this is the ratio of stalk lengths to filter sizes found in filter-feeding protozoa, ranging from 10 to 20 $\mu$m-tall choanoflagellates to nearly 2 mm-high stentors (Figure 3.3).

The discussion of the role of dissolved organics in the nutrition of aquatic invertebrates, and by implication protozoa, has persisted for about fifty years; this literature has most recently been reviewed by Jørgensen (1976). Much of the work attempting to show the significance of dissolved organics in the nutrition of animals is uncritical; in part this is due to a misunderstanding of the significance of the half-saturation constant of the Michaelis-Menten equation as discussed above.

It is a fact that protozoa (and the epithelial cells of many inverte-brates) show both an active uptake of low-molecular weight compounds and an uptake of dissolved macromolecules through "pinocytosis." Some smaller protozoa can be grown axenically in solutions of substrates like peptone or even in chemically defined media. The mouthless mutants of *Tetrahymena* studied by Rasmussen & Orias (1975) constitute a very convincing proof of this. Nevertheless, a number of arguments render it very improbable that dissolved nutrients play any role for free-living protozoa in nature.

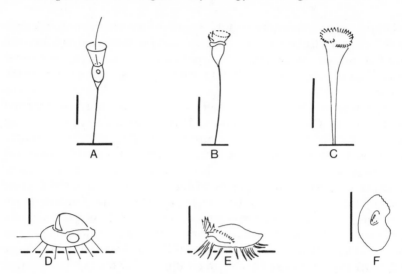

**Figure 3.3**  Ratio between distance of filter from a solid substrate and the radius of the filter in three filter-feeding protozoa which generate water currents perpendicular to the substrate (A-C) and two species (D,E) which generate water currents parallel to the substrate. In the former case this ratio is around 8, in the latter case 3–4. (A: the choanoflagellate *Salpingoeca*, B: *Vorticella*, C: *Stentor*, D: *Cyclidium*, E: *Euplotes*.) F shows the small mouth of filter feeders living in narrow crevices (*Colpoda*). Scale bars: A: 10 μm, B: 100 μm, C: 1 mm, D: 10 μm, E: 50 μm, F: 50 μm.

The most important argument in favor of this view is, perhaps, that no protozoan groups have evolved free-living (non-photosynthetic) forms which do not ingest particulate material. A few parasitic forms (e.g., the astome ciliates living in the intestine of annelid worms) must depend on dissolved nutrients, but even the majority of parasitic protozoa are phagotrophs. A few other protists (e.g., apochlorotic chlamydomonad flagellates and diatoms) which live in special environments with high concentrations of dissolved organic material exist, but this is also very rare.

The reason for this is probably that the relatively larger protozoa cannot compete with bacteria for dissolved nutrients. The fact that the specific clearance of a diffusion feeder is inversely proportional to the square of its length suggests that large cells will be less efficient than small cells in competing for dissolved nutrients. In order to grow *Tetrahymena* cells axenically, peptone solutions of 2 to 5 g per liter of medium are necessary (see Rasmussen & Orias, 1975). These ciliates will, however, grow at approximately the same rate in a suspension of about $5 \times 10^6$ bacteria/ml which is only about $5 \times 10^{-4}$ times the concentration of organic material needed for growth based on dissolved nutrients. Comparisons of the volume-specific clearance for amino acids by *Escherichia coli* reported by Koch (1971) are about a thousand-fold higher than similar values for amino acid uptake in *Tetrahymena*, as calculated from the data in Jørgensen (1976). In this context the observation that the half-saturation constants (K) for amino acid uptake are comparable (within the range of 1 to 8 $\mu$M for the bacterium and within the range 8 to 94 $\mu$M in the case of the ciliate) is misleading. Since K is proportional to the maximum rate of uptake, Vm, this observation only reflects the fact that the Vm per unit volume of ciliates is very low as compared to similar values for bacteria.

It is therefore unlikely that the uptake of dissolved organic matter or that phenomena such as pinocytosis in large amoebae (Chapman-Andresen, 1967) play any significant nutritive role in free-living protozoa. The real function of these processes in protozoa (and in epithelial cells of invertebrates) is not well understood. It may be possible that their primary function is related to the maintenance of the steep chemical gradients of dissolved organic compounds which occur across the cell membrane.

## 3.2   Feeding in true protozoa

In spite of the diversity of feeding mechanisms, it is still possible to classify protozoa according to three categories, namely, filter feeders, raptorial feeders, and diffusion feeders. This classification will be used as a framework for the following discussion.

**Filter feeders**     Filter feeding is an adaptation for feeding on small, suspended food particles. In filter-feeding flagellates, a flagellum propels water through a collar of nearly straight and seemingly rigid tentacles or pseudopodia which act as a filter. In the choanoflagellates, the collar typically consists of 20 to 50 tentacles, each about 0.1 $\mu$m thick. The free distance between neighboring tentacles is only 0.1 to 0.3 $\mu$m. In the center of the base of the collar one smooth flagellum is situated; it drives water away from the cell so that suspended particles are collected on the outside of the collar. The food particles are then phagocytized by pseudopodia arising from the periphery of the collar (Fenchel, 1982a; Laval, 1971; Leadbeater & Morton, 1974; see also Figures 3.4,A; 8.2,A,C,E; and 9.1A). The choanoflagellates are specialized for feeding on the smallest prokaryotic cells so they play an important role in planktonic environments. The low porosity of the filter explains the slow water currents through it (10–20 $\mu$m/sec), but due to the small size of choanoflagellates and thus their large surface to volume ratio,

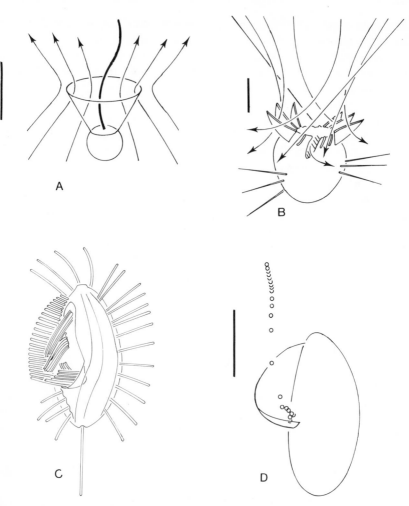

**Figure 3.4**   Flow lines generated by protozoan filter feeders. A. The choano-flagellate *Diaphanoeca*; food particles are intercepted on the outside of the tentacular collar. B. The oligotrich ciliate *Halteria*; food particles are inter-cepted on the inside of the membranelle zone. C,D. The scuticociliate *Cycli-dium*: food particles are intercepted on the paroral membrane which consists of a row of immobile, parallel cilia. In D the position of a latex bead was recorded every 20 msec on a video recorder; it shows how water is accelerated by the oral membranelles to reach velocities of around 300 μm per second and how the particle is eventually intercepted by the ciliary membrane. All scale bars are 10 μm. (Redrawn from Fenchel, 1986a.)

they filter about $10^5$ times their cell volume of water per hour, equivalent to 1 to $4 \times 10^{-6}$ ml/h (Fenchel, 1986a).

Certain chrysomonads and helioflagellates, although entirely unrelated to the choanoflagellates, apply a similar system for catching bacteria. However, these forms have hairy flagella, and so in this case water current is directed against the cell and the food particles are intercepted along the inside of the tentacle collar. These forms have a much coarser filter and generate much faster water currents than do the choanoflagellates. Due to the simple geometry of filter-feeding flagellates, and in particular their radial symmetry, it has been possible to make rather detailed hydrodynamic models of the feeding currents and to compare them with the flow fields and filtration rates of the real organisms (Fenchel, 1986a; Lighthill, 1976).

It is among the ciliates that we find the greatest diversity and specialization for filter feeding. Filter feeding is often found among the oligohymenophorans (hymenostomes such as *Tetrahymena, Colpidium*, and *Paramecium*, scuticociliates such as *Cyclidium* and the peritrichs), among the colpodis, and among the polyhymenophorans (hypotrichs such as *Euplotes*, heterotrichs such as *Stentor* and *Blepharisma*, and oligotrichs such as *Halteria* and the planktonic tintinnids).

In filter-feeding ciliates, the water currents are always generated by membranelle zones which are situated on the left side of the mouth (Section 2.2). However, there are two different principles of filtration. In some forms the membranelle zone not only generates the water currents, but also functions as a filter. The membranelles pump water out of the mouth, while particles which are too large to pass between neighboring membranelles are retained along the inside of the zone and eventually accumulate at the cytostome. Such "upstream" filtration is typical of all the polyhymenophoran ciliates (Figures 2.1 and 3.4,B) and the colpodids (to the extent that these have been studied); it is also found in a few oligohymenopheran ciliates. High water velocities (0.5—1 mm/ sec) are characteristic for this type of particle retention. The lower part of the size spectrum of retained particles is also ill-defined (Figure 3.5). This is because the distance between neighboring membranelles varies during the beat cycle. The mechanism does not allow the retention of very small particles; the species with the tightest arrangement of membranelles will retain 100 percent of all particles exceeding 1—2 $\mu$m.

In most oligohymenophoran ciliates, the water currents generated by the membranelles are driven through an array of parallel and nearly motionless cilia on the right side of the mouth, the "paroral" or "undulating membrane," which acts as a sieve. A characteristic example is *Cyclidium* (Figure 3.4,C,D). In peritrich ciliates, one membranelle and the paroral membrane run counter-clockwise all the way around the pole of the cell and then enter the funnel-shaped "infundibulum" lead-

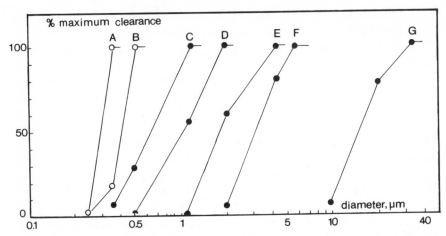

**Figure 3.5** The retention efficiencies for the lower size range of intercepted particles in seven filter-feeding ciliates. Open circles: particle retention with a paroral membrane; filled circles: retention with a membranelle zone. A: *Cyclidium citrullus*; B: *Colpidium colpoda*; C: *Uronema marinum*; D: *Halteria grandinella*; E: *Euplotes moebiusi*; F: *Blepharisma japonicum*; G: *Bursaria truncatella*. (After Fenchel, 1986a.)

ing to the cytosome. The undulating motion of the membranelle drives water in between it and the paroral membrane and toward the infundibulum. As the water moves through this passage, most of the water seeps out between the cilia of the membrane, so a concentrated suspension of particles enters the infundibulum. The organisms which use this form of filtration are specialized to feed on relatively small particles (typically as small as 0.3 μm). The size spectra of retained particles show a sharply defined lower-size range in accordance with the fact that the cilia of the undulating membrane are held parallel and nearly motionless during the filtration (Figure 3.5). The water currents generated by the membranelles have velocities comparable to those of upstream filterers. However, the water current approaches the paroral membrane at an acute angle so that the flow velocity through the membrane is quite low, typically 20 μm/sec. Consequently, in many of these forms, the filter area is huge (e.g., *Cyclidium* and the peritrichs).

Some filter-feeding ciliates, however, have quite a small mouth and so have a low clearance. This applies to soil ciliates, such as *Colpoda* (Figure 10.2,A), and also to tetrahymenine ciliates, such as *Colpidium* and *Glaucoma*. This is probably an adaptation to life in small crevices (e.g., soil, detritus, and carrion). As discussed in the preceding section, the hydrodynamic resistance to flow due to the proximity of solid surfaces is a function of the transectional area of the water flow which is filtered. When a ciliate is sufficiently close to a surface there is no gain in clearance by increasing the filter area because this would be counteracted by a corresponding decrease in the average flow velocity through it (Fenchel, 1986a).

**Raptorial feeding**   Raptorial feeding is found in many small flagellates. These and the small amoebae are the only protozoa which are sufficiently small to feed on bacteria in this way. Chrysomonads and the possibly related bicoecids drive water currents against the cell with their anterior, hairy flagellum. Particles which touch a lip-like structure supported by bundles of microtubules and surrounding the cytostome are phagocytized (Figure 3.6,A). The free-living kinetoplastid flagellates are also raptorial feeders. Some forms, such as *Pleuromonas jaculans*, feed on suspended bacteria much as the chrysomonads do. Most kinetoplastids, however, are specialized on bacteria associated with solid surfaces. These flagellates slide on their trailing flagellum and ingest particles which come in contact with the cytostome. The characteristic structure of phagotrophic kinetoplastids are microtubular rods, which form a "cytopharynx" and which must play a role in the ingestion of

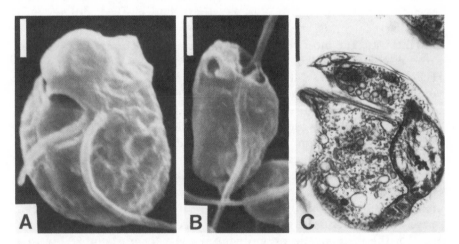

**Figure 3.6**   A. The chrysomonad flagellate *Ochromonas* fixed (with a mixture of $OsO_4$ and $HgCl_2$ solutions) immediately after having ingested a bacterium (the bulge above the closing cytostome). The short and the long flagellum are also seen. B. The kinetoplastid flagellate, *Pleuromonas jaculans*, showing the cytostome enforced by microtubules as well as its two flagella arising in a flagellar pocket. C. A transmission-electron micrograph of *Pleuromonas* showing the microtubules of the cytostome which must play a role in the transport of captured bacteria into food vacuoles. Also seen (to the right) is the swollen, DNA-containing part of the mitochondrion (the "kinetoplast"). Scale bars: A,B: 1 $\mu$m; C: 0.5 $\mu$m.

the food particles (Figure 3.6,B,C; Fenchel, 1982a). Similar rods are also found in the phagotrophic euglenids; *Peranema* and *Entosiphon* (Mignot, 1966; Nisbet, 1974), which also find their food particles along surfaces. Many dinoflagellates are obligatory or facultative predators of other organisms. Phagocytosis takes place in the "sulcus" close to the place where the flagella are attached. In some dinoflagellates (e.g., *Gymnodinium* an extensible structure, the peduncle, which protrudes from the sulcus area, is used for the capture of prey cells and for the ingestion of their cytoplasmic contents (Spero, 1982). In *Gyrodinium* and some of its relatives, exploding trichocysts immobilize the prey organism prior to phagocytosis (Biecheler, 1952). Among the heterotrophic flagellates, *Oxyrrhis* and *Noctiluca* (Figure 8.4,K) have been especially well studied (Droop, 1966; Prasad, 1958). The latter, a giant dinoflagellate, feeds on algal cells, other protozoa, and even crustacean larvae and other small invertebrates. Prey organisms stick to the motile tentacle of the flagellate and are then brought to the cell surface and phagocytized.

Filter feeders retain and ingest all properly sized particles. Raptorial protozoa, on the other hand, which capture each particle individually, rather than in bulk, may show a considerable degree of discrimination on the basis of qualities other than size. Among the ciliates there are several examples of this. It is especially the "primitive" ciliates, that is, the karyorelectids, the prostomatids, and the pleurostomatids which feed this way. The most common prey item among karyorelectids seems to be microalgae and flagellates, but other ciliates are also ingested (Fenchel, 1968). Prey size plays a considerable role in determining whether a particle is ingested or not, and there is a close correlation between protozoan cell size and the preferred particle size (Fenchel, 1968; Finlay & Berninger, 1984; see also Figure 7.1).

The prostomatids comprise some of the best-known protozoan predators. The classical studies of Gause (1934) on the kinetics of a prey-predator system was based on *Didinium nasutum*, which specializes on the use of *Paramecium* as food. This system has been reexamined more recently (Luckinbill, 1973, 1974; Salt, 1979), mainly with emphasis on the functional response of this predator and the stability properties of the system. The prey organism has a length comparable to a *Didinium* cell and the ingestion process, after the predator has immobilized its prey is beautifully illustrated in Wessenberg & Antipa (1970). Like other predatory prostomatids and pleurostomatids, *Didinium* has "toxicysts," extrusomes which explode when they are touched by the prey cell and which subsequently immobilize it. Other common predators of ciliates include *Lacrymaria, Homalozoon, Dileptus* (which may prey on small metazoa as well), *Litonotus*, and *Loxophyllum* (Dragesco, 1962, 1963; Fenchel, 1968; Kuhlmann et al., 1980; see also

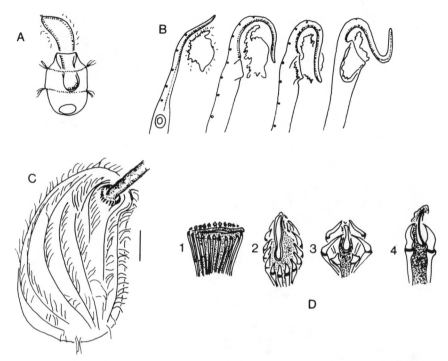

**Figure 3.7** A. A *Didinium nasutum* cell ingesting a *Paramecium* cell. B. The anterior end of *Dileptus anser* ingesting a *Colpidium* cell. Toxicysts situated in the trunk lyse the prey cell which is then ingested. C. *Pseudomicrothorax dubius* using its pharyngeal (microtubular) rods to ingest an *Oscillatoria* filament (scale bar: 10 μm). D. The pharyngeal baskets or rods of some dysteriid ciliates; the variety in structure suggests specializations for different types of food. (1 and 2, redrawn from Dragesco, 1962; 3, redrawn from Peck, 1985; 4, redrawn from Deroux, 1976.)

Figure 3.7,A,B). Some of these forms feed predominantly on colonial, peritrich ciliates (Canella, 1951).

Predators also are found among other ciliate groups. Some hymenostomes, such as *Lembadion* (see Figure 9.2,D), have secondarily evolved raptorial feeding, as has the hypotrich *Uronychia*. *Frontonia*, which is a close relative to *Paramecium*, lacks the filter-feeding habits of its relatives and feeds predaciously on dinoflagellates, large diatoms, and cyanobacteria (Figure 3.8). The hypostome ciliates are characterized by a cytopharyngeal basket of microtubular rods which assists in the ingestion of filamentous organisms (Tucker, 1968). In *Nassula* and its relatives, this organelle is used for feeding on filamentous cyanobacteria on which they have specialized (Dragesco, 1962; Tucker, 1968). The chlamydodontids and the dysteriids feed primarily on bacteria and fila-

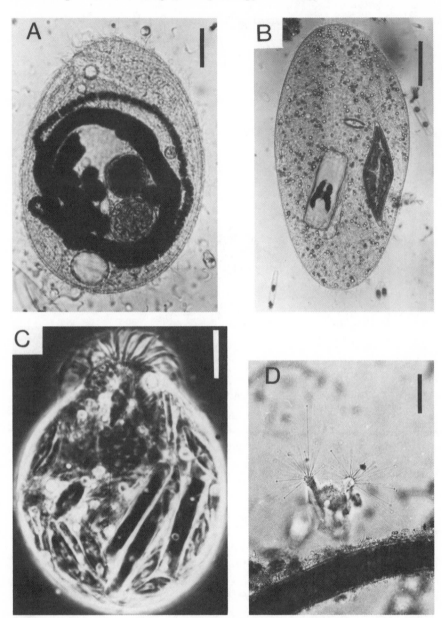

**Figure 3.8**  A. The ciliate *Sonderia schizostoma* containing filaments of the white sulfur bacterium *Beggiatoa*. B. *Frontonia marina* with ingested diatoms. C. *Strombidium* sp. with ingested diatoms. D. The marine suctorian *Acineta tuberosa*. Scale bars: A: 20 μm, B: 50 μm, C: 10 μm, D: 50 μm. (A,B, from Fenchel, 1968.)

mentous microorganisms associated with surfaces. The diversity of the structure of the pharyngeal rods found among these forms suggests a high degree of specialization for specific foods (Deroux, 1976; Peck, 1985; see also Figure 3.7,C,D).

Histophagy is a specialized type of raptorial feeding which has developed independently in several different ciliate taxa. Histophagous ciliates attack damaged, but live, animals such as annelid worms and small crustaceans. The ciliates enter wounds, to which they are attracted by a chemosensory mechanism, and ingest animal tissue. If many ciliates attack simultaneously, which is often the case, everything but the cuticle may be devoured within one hour. The most specialized histophages have a polymorphic life cycle including a swarmer stage, a feeding stage, and a cyst stage (Section 5.1). In these forms, the trophic stage is short-lived (30 to 60 minutes) and during this period the ciliates may ingest ten times their own volume. Histophagy occurs among some prostomatids (e.g., *Coleps* and *Prorodon*) where it has evolved as a result of the carnivorous habits of these ciliates. In other groups, histophagous species evolved from forms which feed on bacteria in carrion (e.g., some species of *Tetrahymena* and some philasterid ciliates). The most specialized histophages are found in the genus *Ophryoglena* (Figure 3.9). The biology of histophagous ciliates has been the subject of several studies (e.g., Canella & Rocchi-Canella, 1976; Corliss, 1973; Fenchel, 1968; Savoie, 1968).

Among the sarcodines, raptorial feeding is especially common in lobose amoebae. The feeding of large amoebae on ciliates and the method by which amoebae surround the ciliates using pseudopodia and enclose them in a "food cup" prior to food vacuole formation is well documented and has probably been observed by most biologists (Lindberg & Bovee, 1976). The numerous species of small marine, freshwater, and soil amoebae depend primarily on bacteria and also on microalgae for food (Page, 1976, 1983; Schuster, 1979). This also applies to most lobose and filose testaceans (Heal, 1961; Ogden & Hedley, 1980; Stump, 1935).

**Diffusion feeding**  Diffusion feeding is found among many types of sarcodines. In heliozoa, axopods radiate out in all directions (Figure 2.3). Prey cells which happen to collide with the axopods are held by mechanisms which are not fully understood but which involve special extrusomes. The prey may then be brought closer to the cell through the bending of the axopods or through cytoplasmic streaming on the surface of the axopods. Eventually a pseudopodium arising from a cell surface engulfs the prey and forms a food vacuole. The largest heliozoa (*Actinosphaerium*) feed on large ciliates and on small metazoa, such as rotifers and crustaceans, while the smallest forms depend on motile

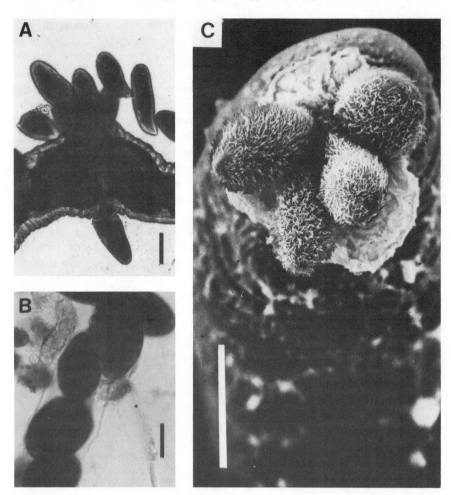

**Figure 3.9** The marine histophagous ciliate *Ophryoglena macrostoma*. A. Tomites entering a cut in an oligochaete worm. B. The same preparation after 45 minutes showing the trophonts within the empty worm-cuticle. C. Scanning-electron micrograph of young trophonts feeding on a cut oligochaete. All scale bars: 100 μm.

bacteria and small flagellates. Heliozoan feeding has been described in detail by Dragesco (1964) and by Hausmann & Patterson (1982).

Feeding in planktonic foraminifera was studied by Anderson & Bé (1976a) and Anderson et al. (1979). They examined the process of feeding of the large (1 mm) *Hastigerina* on larvae of the crustacean *Artemia*. In this case the larval prey stick to the reticulopodial network, which eventually envelopes it and then penetrates it. The prey is then

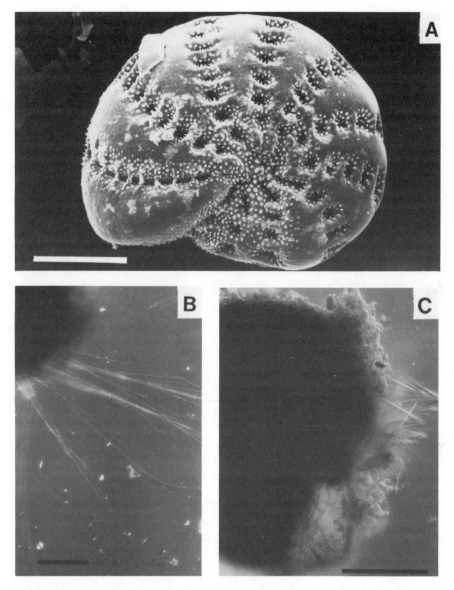

**Figure 3.10**   The foraminifer *Elphidium williamsoni*, showing the test (A), the reticulopodia which trap food particles (B), and the feeding cyst (C). All scale bars: 100 μm.

drawn into the ectoplasmic "bubble layer" where digestion takes place in a food vacuole. In radiolaria, axopodia or reticulopodia hold on to any prey organisms which collide with them. Some radiolaria are also surrounded by a "sticky jelly" which helps retain prey, while in small species the spines may assist mechanically as the reticulopodia are anchored to them as well as to the struggling prey. Radiolarian species of different sizes prey on small metazoa, ciliates, and algae (Anderson, 1983; Swanberg & Anderson, 1985).

In contrast to the organisms discussed above, most Granuloreticulosa and thus the majority of the foraminifera are associated with solid surfaces over which they spread a network of often extremely thin reticulopodia (Figure 3.10). Prey organisms, such as algal cells, bacteria, or other small cells, stick to the pseudopodia and are drawn toward the predator by contraction or cytoplasmic streaming. Many foraminifera are surrounded by empty diatom frustules and other remains, the contents of which were apparently digested in food vacuoles outside the test; in other cases particulate material is brought inside the test. Detailed studies on the feeding and food requirements of mainly estuarine species have been carried out by Lee (1980); Lee et al. (1969); Lee & Muller (1973), and Muller (1975). The study by Christiansen (1971) demonstrates a diversity and inventiveness among some benthic foraminifera in their use of pseudopodial nets for the capture of different kinds of prey (see also Figure 8.9).

Among the ciliates, the suctorians make use of diffusion feeding and their prey organisms are almost exclusively other ciliates. Mature suctorians are devoid of cilia and in most cases are attached (with or without a stalk) to a substrate. They possess bundles of tentacles which are supported by an internal cylinder of microtubules (Figure 3.8,D). Extrusomes on the tip of the tentacles serve to attach and immobilze any ciliates which touch them. The tentacles eventually penetrate the cell membrane of the prey, whose contents are then drawn through the tentacle and into the suctorian predator. (Some suctorians have adopted a parasitic life, living intracellularly in ciliates which are much larger than themselves.) Bardele & Grell (1967) and Bardele (1972) described the adhesion of the prey and suggested a mechanical model for the movement of prey cytoplasm through the tentacle. A general review of the biology of suctoria appears in a paper by Canella (1957). The haptorid ciliates belonging to the genus *Actinobolina* (Figure 9.2,C) also use tentacles to catch swimming prey organisms such as rotifers. Their tentacles carry toxicysts which immobilize the prey which is subsequently ingested through the cytostome.

# 4

# Ecological Physiology: Bioenergetics

## 4.1   Balanced growth: the efficiency of conversion

In order to grow and divide, protozoa must assimilate building blocks for the synthesis of cell constituents (assimilatory metabolism). They also need free energy for the synthesis of macromolecules, for the maintenance of the integrity of the cell, and for various processes which enhance the survival of the cell (dissimilatory metabolism). Energy is also needed for osmotic and electric work and for motility. In heterotrophic eukaryotes, the assimilatory as well as the dissimilatory metabolism is based on organic molecules. In an ecological context it is of interest to know the efficiency of conversion of food to cell biomass and the rate at which this conversion takes place.

In most treatments of ecological bioenergetics, "balanced growth" is implicitly assumed. In a constant environment a cell population will increase exponentially and irrespective of how we quantify the population (e.g., cell numbers, organic carbon, DNA, oxygen consumption) the growth rate constant will remain the same. Balanced growth implies that the "age structure" (that is, the relative abundance of the different life cycle stages) is constant over time. In a constant environment this will eventually be the case even if the population initially has a special structure, for example, if the growth cycles of the individual cells are synchronized. Balanced growth is a property of the population, not of individual cells, in which different processes, such as DNA synthesis, take place during certain stages of the cell growth cycle.

In principle, at least, balanced growth occurs in a chemostat. Such growth can also be more or less attained in a batch culture, if the initial

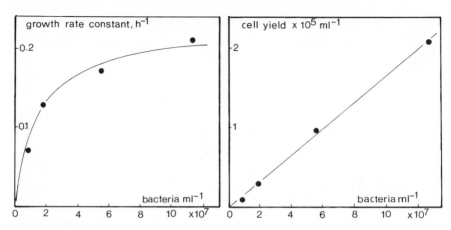

**Figure 4.1**    (Left) the growth rate constant (at 20°C) of the phagotrophic chrysomonad *Paraphysomonas vestita*, as a function of environmental concentration of bacteria. (Right) the eventual yield of flagellates as a function of the initial number of bacteria in four batch cultures. (Data from Fenchel, 1982b.)

inoculum is very small so that the population can grow for several generations without changing the environment much. In nature, if balanced growth is realized at all, it only occurs over very short periods. Although it is a useful concept, balanced growth is an abstraction which can only be approximated under experimental conditions.

When expressed in units of energy, the ingested material is equal to the sum of respiration + growth + egested material + excreted material. Respiration represents the dissimilatory metabolism; in practice, it is usually measured as $CO_2$ production or $O_2$ consumption. Oxygen consumption (in aerobic forms) is an adequate measure of power generation (about $2 \times 10^8$ erg or 20 J per ml $O_2$, see Fenchel & Finlay, 1983). Egested and excreted materials represent nondigestable parts of the food and loss of dissolved low-molecular weight compounds respectively; in practice, they are difficult to measure separately. The "gross growth efficiency" or "yield" is defined as growth divided by ingestion; "net growth efficiency" is growth divided by (respiration + growth). Yield can be measured in a batch culture as the eventual yield of cells divided by the initial amount of food particles added to the culture. (In practice, however, certain corrections must be made because during the last stages of a batch culture, growth is not balanced and the average cell size decreases. This difficulty does not arise if the measurement is made in a chemostat.) If respiration is also measured, net growth efficiency can be determined.

Empirical measurements show that growth efficiency is nearly invariant with growth rate in protozoa (Figure 4.1). This finding will not appear unusual to a microbiologist, but it is strange to a zoologist (since in larger animals a large fraction of the energy budget is spent on maintenance and is invariant with growth). It has several important implications: One is that the growth rate constant is almost linearly proportional to food consumption, using the yield as the proportionality constant. The growth response of a population to food particle density therefore has the same functional form as that of the ingestion rate (Figure 4.1, see also Section 3.1). It is also evident that the respiration rate must be linearly related to the growth rate constant within the range where balanced growth can be maintained. The concept of "basal metabolism," as used by zoophysiologists, has therefore little meaning in the case of protozoa. It reflects the fact that in small organisms by far the largest part of the power generation is involved in macromolecular synthesis and thus is directly coupled to growth. Conversely, only a very small fraction of the energy budget is spent on mechanical, electrical, or osmotic work.

It is illuminating to calculate the approximate cost of motility. As an example, consider a spherical flagellate (radius: 4 $\mu$m) which swims with a velocity of 60 $\mu$m/sec (these figures fit an *Ochromonas* cell closely). From Stokes' law we find that the necessary power is $6\pi r v^2 \eta$, where v is velocity and $\eta$ is viscosity (0.01 poise). In the case of the above flagellate the required power equals $2.7 \times 10^{-9}$ erg/sec. If we assume that the overall efficiency (hydrodynamic efficiency times the efficiency of converting chemical work into mechanical work) is 1 percent, the total power required is $2.7 \times 10^{-7}$ erg/sec. This estimate agrees with independently estimated values for the power consumption of a single flagellum of $2-8 \times 10^{-7}$ erg/sec (Sleigh, 1974). A growing flagellate of this size consumes about $4.5 \times 10^{-3}$ nl $O_2$ cell/h which corresponds to a power generation of $2.5 \times 10^{-4}$ erg/sec. Therefore only about 0.1 percent of the energy budget of the flagellate is used for motility; this figure increases somewhat for a starving cell in which the respiratory rate may be considerably reduced. This and similar examples (Fenchel, 1986a; Fenchel and Finlay, 1983) show that protozoa in general use only a small fraction of their power generation for motility. This is in contrast to larger animals in which motility constitutes a substantial part of the energy budget and in which an increase in motility leads to a significant increase in the respiratory rate. There are two reasons for this difference. One is a function of the fact that the volume-specific respiration of small organisms is very high relative to larger organisms. The other is related to the fact that the energetic costs of the slow swimming characteristic of protozoa are very low. Similar considerations show that the power needed for osmotic and electric work is also modest as compared to the respiratory rate of protozoa.

For bacteria it has been shown that during balanced growth of carbon-limited cells the generation of one mole of ATP leads to the production of approximately 10 g dry weight of cells (Beauchop & Elsden, 1960; Paine, 1970). If (as is nearly always the case for eukaryotic cells) the C atoms of the substrate have an oxidation level similar to that of glucose, this corresponds to a net growth efficiency of 67 percent. In unicellular eukaryotes (and in growing tissue of poikilotherm metazoa) it has also been found that net growth efficiency is around 60 percent (Calow, 1977; Fenchel & Finlay, 1983). This value is invariant with size of the organism and (within wide limits) with growth rate and probably represents some fundamental limit to growth efficiency. Gross growth efficiency is a more variable parameter; typical values quoted in the literature are 30 to 50 percent so that as much as 50 percent of the ingested material may be egested in an undigested state or is lost as dissolved organic material. Gross growth efficiency probably varies not only among different species, but also according to the qualitative nature of the food particles (Fenchel, 1982b; Fenchel & Finlay, 1983, and references therein).

## 4.2    Balanced growth: the rate of living

All species studied in detail can perform balanced growth over a range of rate constants. These depend, of course, on temperature, but also at a given temperature (within a certain physiological range) the cells can grow at varying rates depending on the quantity and quality of the available food particles (Figure 4.1). Curiously, this fact is often considered trivial. As pointed out by Koch (1971), changes in growth rates require complex adjustments in the rates of synthesis of macromolecules and must be considered an adaptive trait. When a given range in growth rate constants of an organism has been established in the laboratory, one can be fairly certain that this entire range is also realized in nature at one time or another.

A close approximation to theoretical maximum growth rate can be realized in laboratory cultures of many species. In small flagellates and ciliates, minimum doubling times of 2.5 to 3 hours (20°C) are found, in larger ciliates the corresponding figures are 10 to 20 hours and in large sarcodines, several days. There is also a lower threshold of food intake below which growth ceases and the cell changes to a physiological state adapted to the endurance of starvation (Fenchel, 1982c; Trinci & Thurston, 1976). It is, however, very difficult to maintain balanced growth (or any reasonable approximation) with very low rate constants under experimental conditions. This applies to batch as well as to chemostat cultures and the problem of a minimum growth rate has attracted little attention. Fenchel (1982b) found that a heterotrophic flagellate can maintain balanced growth within a range of doubling times from about 3 hours up to around 24 hours, but the latter value is rather inaccurate. Without more data it is not clear whether a range of potential growth rates spanning a factor of about ten is typical. This range may vary considerably among species depending on the variation in resource availability which is characteristic of their habitats.

It has long been known that the "rate of living" expressed in terms of growth rate or weight-specific metabolism tends to decrease with body size when organisms covering a large size span are compared. Very crudely, these rates are inversely proportional to body length. This has been rationalized from the surface:volume ratio. In homoitherm organisms the heat loss across the surface would control the metabolic rate, but the argument can also be extended to poikilotherm organisms since rates of biological processes could be a function of transport across surfaces. It would then be expected that weight-specific respiration and growth rate are proportional to the one-third power of weight ("Rubner's law"). With respect to respiratory rate, extensive data show that the exponent is actually closer to one-fourth (Hemmingsen, 1960; Zeuthen, 1953); that is, specific metabolic rate decreases somewhat more slowly as a function of weight than suggested by the "surface law." Many attempts to rationalize this (recently reviewed in Schmidt-Nielsen, 1984) have been largely unsuccessful. Nevertheless, it would seem that the idea of a surface relationship does contain a fundamental truth and empirical evidence suggests that during the ontogeny of individual species, Rubner's law holds, but that some other principle must play a role when a large range of sizes of species are compared. Describing weight-specific metabolic rate with the equation, $R = aW^b$, Hemmingsen (1960) interpreted the data so that for both poikilotherm metazoa and protozoa the exponent, b, equals 0.25, while the constant, a, has a value which is seven times higher in metazoa than in protozoa. Zeuthen (1953) believed this was due to the fact that an increase in body size in protozoa means an increase in cell size. Conversely, metazoa are colonies of many relatively small cells whose size depend on cell numbers. Therefore, comparing a small metazoan organism with an equally large protozoan, the former would have a much larger cellular surface area and consequently a higher metabolic rate.

Fenchel & Finlay (1983), who reviewed all published data on protozoan respiration, found that the data show a tremendous scatter, with values varying by a factor of 10–20 for similar sized (and even identical) organisms. The reason for this is that most investigators have not considered the physiological state of the cells during their measurements. As we have seen, respiratory rate is closely coupled to growth in protozoa. A starved cell rapidly decreases its metabolic rate, sometimes by

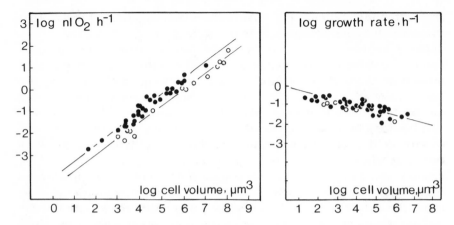

**Figure 4.2** (Left) respiration rates (at 20°C) of various species of ciliates and flagellates (filled circles) and of amoebae (open circles) during exponential growth and as a function of cell volume. The slope of the regression lines is 0.75. (Right) the maximum attainable growth rate constant (20°C) for some protozoan species; the slope of the line is −0.25. (Data from Fenchel & Finlay, 1983.)

a factor of ten or more. This may occur within the time typically used for obtaining respiration measurements of protozoa. A protozoan with a generation time of three hours and a net growth efficiency of 60 percent will metabolize the entire cell-carbon content in 4.5 hours. When placed in a respirometer and left without food for several hours the cell survives, but only because of a rapid reduction in metabolic rate. This has been overlooked in many determinations of respiration rate. A meaningful comparison of protozoan respiratory rates can only be made if the physiological state of the cell is defined. Considering only cells growing at approximately their maximum rate (Figure 4.2) and comparing cells over a size range of six orders of magnitude, the data conform closely to the allometric relationship, $R = aW^b$, in which b is not significantly different from −0.25. For ciliates and flagellates, the value of a is identical to that for metazoa as reported by Hemmingsen (1960). This suggests that there is, in fact, no discontinuity in metabolic rate when uni—and multicellular organisms of similar sizes are compared. Sarcodines, however, do seem to have systematically lower metabolic rates when compared to flagellates and ciliates of the same size.

The maximum attainable growth rates for protozoan species of diverse sizes were compared by Fenchel (1974) and Fenchel & Finlay (1983). As expected, the maximum growth rate constant decreases with about the 0.25 power of cell volume, while, in accordance with the data on respiratory rates, sarcodines seem to have a systematically lower maximum growth rate constant (Figure 4.2).

The conclusions which can be drawn from this sort of comparison, based on double logarithmic presentation and covering many orders of magnitude, have limitations. The graphs give a general picture of the potential rate of turnover of protozoan populations and its variation with cell size. The precise values of the constants describing the regression lines should be interpreted with caution, since there are many sources of error in the individual data points, especially concerning estimates of cell volume.

The regression lines, which describe growth and respiratory rates as a function of cell size, presumably represent some fundamental physiological limit. The reason that they take this particular functional form and the basis for the given values of the constants is not known. However, most of the organisms studied are species which, in nature, depend on patchy occurrences of high densities of food particles. It is therefore likely that the ability to achieve the maximum possible growth rate is an adaptive trait. There may well exist species adapted to environments where food resources are more homogeneously distributed in time and space and for which there is no selective advantage for very rapid growth beyond that which the food concentration can support. It is likely that such species would exhibit a smaller range in potential growth rates and a lower maximum growth rate than predicted from the regression line in Figure 4.2. Among bacteria, such a spectrum of potential growth rates has been established (Veldkamp & Jannasch, 1972), but among protozoa the available data are still insufficient.

## 4.3 Nonbalanced growth: a feast and famine existence

In nature, the availability of food resources is quite patchy in time and space and microorganisms must adapt to this by changing their physiological state. At the molecular level, this has been studied in detail in prokaryotic cells (see Koch, 1971, who also used the expression "feast and famine existence" to describe the conditions for life for *Escherichia coli*). Such detailed studies do not exist for unicellular eukaryotes, but the phenomena which are directly observable when cells are suddenly exposed to an increase or decrease in food resources ("shift up" and "shift down" experiments) are comparable to that observed in bacteria. Following the change there is a "lag time" during which macromolecular synthesis, respiration rates, and cell volume change, until eventually the growth rate constant reaches a level corresponding to the new conditions (Cameron, 1973). As in bacteria,

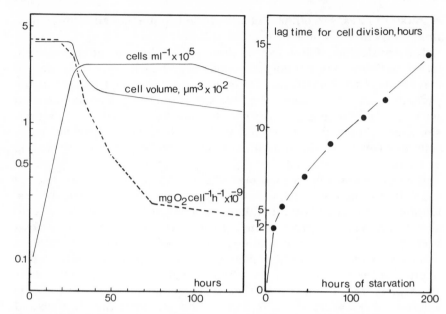

**Figure 4.3** Cell numbers, respiration per cell, and cell volume in a batch culture of the phagotrophic flagellate *Ochromonas* during exponential growth and after onset of starvation (all food bacteria have been consumed) at around 30 hours (left). To the right, the lag time before starving cells begin to divide following feeding as a function of the length of the starvation period. $T_2$ is the generation time corresponding to the food concentration. (After Fenchel, 1982c.)

during balanced growth cell volume increases with the growth rate; this presumably reflects the need for more cell organelles such as ribosomes and mitochondria during rapid growth.

When growing cells are exposed to sufficiently low food levels or are starved, a number of more drastic physiological changes take place. Fenchel (1982c), Finlay, Span & Ochsenbein-Gattler (1983a), and Nilsson (1970) studied various aspects of starvation in the chrysomonad flagellate *Ochromonas* and in the ciliates *Tetrahymena* and *Paramecium*. None of the studied forms responded to starvation by forming cysts (see Section 5.1). After the onset of starvation, cell division continues for one to two generations, yielding cells of smaller-than-normal size. The respiration rate decreases immediately and in *Ochromonas* eventually falls to 2 to 5 percent of that of growing cells (Figure 4.3). In the starved cells, autophagous vacuoles soon become evident; in *Ochromonas* cells starved for 100 hours, about 80 percent of the cell has been consumed. In particular, mitochondria are digested, so that in cells starved for 80 hours only about 10 percent of the initial mitochondrial volume remains. Neither the loss of mitochondria nor the decrease in the level of enzymes related to the electron-transport chain, however, can totally account for the decrease in respiratory rate; rather, the cells seem to retain the potential for a metabolic rate above that realized during starvation. When the starved cells are fed again, cell volume, the level of electron-transport chain enzymes, and the respiratory rate increases, until eventually the cells start to divide. This lag time before division begins increases for longer starvation periods and eventually becomes several times longer than the generation time corresponding to the food level offered.

The findings of these experiments may be interpreted as an evolutionary compromise between two different fitness requirements of cells exposed to starvation. One goal is staying alive: To do so the organism should completely stop biosynthesis of macromolecules and minimize respiration; an extreme solution which permits this is to encyst forming an almost totally inactive state as some protozoa do. The cost of this, however, is that once food becomes available, the lag time before the resumption of growth increases, so that the organism becomes less competitive. The cells may therefore sacrifice longevity by retaining a certain minimum respiratory level and biosynthetic capacity. The balance between the opposing fitness components is likely to reflect the expected time scale of fluctuations in food resources in the natural environment. If periods of hardship are likely to be short, the maintenance of relatively high physiological activity should be favored; at the other extreme, during long starvation periods cyst formation could be expected. There is considerable scope for comparative studies of these phenomena in protozoan isolates from diverse types of habitats.

## 4.4   Anaerobic metabolism

There is little doubt that aerobic metabolism and the possession of mitochondria is a primary original feature of eukaryotes; also among free-living protozoa, the overwhelming majority are aerobes. Within several groups of protozoa, however, obligatory or facultative anaerobes are found. Best known are intestinal commensals, such as the ciliate fauna of the rumen, the hypermastigid flagellates of the termite hindgut, and a variety of intestinal flagellates of vertebrates. Among free-living forms, there are several heterotrichid and trichostomatid ciliates which occur in anaerobic waters and sediments: Such environments also harbor a number of different flagellates (see Sections 8.3 and 9.4). The diplomonad flagellates, of which some are free-living (*Trepomonas, Hexamita*), do not possess mitochondria (Lee et al., 1985). This is also the case for the above-mentioned ciliates, which, in addition, are sensitive to the presence of oxygen (Fenchel, Perry & Thane, 1977).

Very little is known about the metabolic pathways employed by free-living anaerobic protozoa, but it is reasonable to assume that some resemblance with the better-studied parasitic and commensal forms exists. In all cases, some sort of fermentative process is likely to be responsible for power generation. A variety of fermentative processes with diverse ATP-yields are possible; in all cases, the ATP-yield (and therefore the net growth efficiency) is much lower than in aerobic metabolism. In termite flagellates, rumen ciliates, and trichomonad flagellates, acetate and butyrate seem to be the main end products of fermentation (Coleman, 1979; Hungate, 1955, 1975; Müller, 1980), suggesting a rather efficient, *Clostridium*-type of fermentation. The biochemical basis for this reaction is the removal of reduction equivalents in the form of hydrogen. In trichomonad flagellates, this takes place in a special organelle, the hydrogenosome (Müller, 1980). Electron microscopy shows organelles resembling hydrogenosomes in a variety of other anaerobic protozoa, including free-living ciliates (Fenchel et al., 1977): Thus, the metabolism of these forms probably is comparable to that found in the trichomonad flagellates. The association of ecto—and endosymbiotic bacteria with anaerobic protozoa is probably also related to hydrogen production (see Section 6.3).

At the moment, there is only one example known of a different type of anaerobic metabolism, namely, in the ciliate *Loxodes*. This form possesses mitochondria and does respire oxygen. These ciliates normally occur close to the oxic-anoxic boundary layer of the water column or sediment of lakes and are often exposed to and survive anoxic conditions. When exposed to anoxia, the number of mitochondrial cristae and the amount of enzymes involved in the electron-transfer system increases. The ciliate, which contains the enzyme nitrate reductase, may be able to employ dissimilatory nitrate respiration, which is otherwise only known in prokaryotic cells (Finlay, 1985; Finlay, Span & Harmon, 1983).

# 5

# Ecological Physiology: Other Aspects

## 5.1  Polymorphic life cycles

Between two divisions, a cell will pass through a number of temporally more or less well-defined phases characterized by DNA synthesis, various morphogenetic events, protein synthesis, and growth. During balanced growth the temporal succession of these events is well defined relative to the time of cell division. However, in many protozoa, at least during constant environmental conditions, the phenotype remains fairly constant except for changes in cell size and, of course, except during cell division. In some species, however, the phenotypic appearance may change considerably, either as an obligatory part of the life cycle, or as a result of environmental changes. Cases in which sexuality is involved will be discussed in Section 5.2.

Encystment plays an important role in protozoan ecology and may serve several different purposes. One such role is protection against physically adverse conditions. One universal requirement for protozoan activity is water and species associated with terrestrial or semi-terrestrial environments such as soils and leaf litter are often capable of producing desiccation-resistant cysts in response to the evaporation of the surrounding water. Most attention has been given to soil ciliates within the genus *Colpoda*. Their protective cysts remain viable for many years in a desiccated state, and for shorter periods they can survive dry heat up to 120°C, humid heat up to 50°C, and submersion in liquid nitrogen. These cysts also survive passage through the gut of animals. On leaves, *Colpoda cucullus* is capable of completing its life cycle by excysting in dew drops during the night and encysting when the water evaporates

63

during the day (Dawson & Hewitt, 1931; Mueller & Mueller, 1970, and references therein).

Short time intervals between encysted and active forms are also reported for the marine, oligotrich ciliate *Strombidium oculatum*. Fauré-Fremiet (1948) studied a population in intertidal pools. During low tide the ciliates were swimming actively, but at high tide the ciliates were encysted at the bottom of the pool, which prevented them from being washed out with the tide.

Perhaps the most common stimulus which leads to encystment is starvation. Not all protozoa react in this way, but it is a very widespread phenomenon and is easily observed in cultures of *Didinium* and *Colpoda*, and also in many other ciliates, flagellates, and sarcodines. Some forms such as *Paramecium* never encyst. It seems, although it has not been systematically studied, that polymorphism with respect to cyst formation occurs within some protozoan populations in nature and that laboratory populations often lose this capacity (Fenchel, 1982c).

It is often assumed that the ability to disperse is an advantageous trait and that this, in part, explains the formation of protective cysts. Although desiccation-resistant cysts may be dispersed widely in an airborne state, it is very hard to evaluate whether this *per se* has adaptive significance. Short-distance dispersal of cysts, however, does seem to play a role in at least one case, that of the ciliate *Sorogena*. This most peculiar organism has recently been described in terrestrial environments from several tropical and temperate localities (Bradbury & Olive, 1980; Olive & Blanton, 1980; Figure 5.1). It has been described as a haptorid ciliate related to *Enchelys*, but Foissner (1985a) considers it to be a colpodid ciliate. It lives in water drops on leaves where it feeds on *Colpoda* and possibly other protozoa. When its food supply is exhausted, the *Sorogena* cells congregate into small heaps. The cells then excrete a stalk and eventually form an aerial "sporocarp" resembling that of slime molds. The cell, now encysted, are found on the tip of the stalk in a spherical conglomerate. After desiccation, a slight mechanical disturbance will lead to dispersal of the cysts.

In the literature, there are many reports on other agents which induce encystment or excystment of protozoa, such as changes in pH, ionic composition of the water, temperature, or light. The adaptive significance of many of these findings is unclear. A general review of cyst formation is given by Corliss & Esser (1974).

More complex life cycles are found in soil amoebae of the *Naegleria* and *Tetramitus* types. These occur either as cysts, as amoebae, or in flagellated forms. The amoebal form, which is a trophic stage, encysts in response to desiccation and starvation. If distilled water is added to cultures, a situation which mimics rainfall, the amoebae transform themselves into flagellates, which in the case of *Tetramitus* are also capable

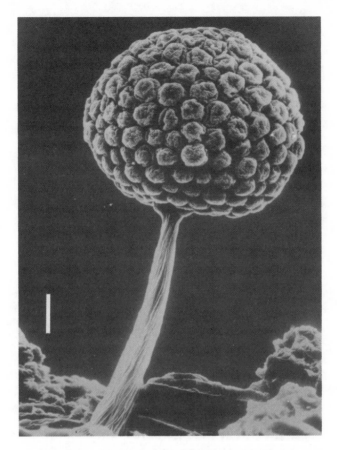

**Figure 5.1** An air-dried and gold-coated sporocarp of the ciliate *Sorogena* sp. seen in the scanning-electron microscope. The cysts survived the treatment and eventually excysted. Scale bar: 20 μm. (Micrograph by Barbara Grimes.)

of feeding on bacteria. To some extent these transformations take place spontaneously in culture and also in response to chemical stimuli, the ecological significance of which are less obvious (Schuster, 1979).

Many ciliates and hetereotrophic flagellates respond to starvation by carrying out one or two further divisions which result in two or four cells with approximately one-half or one-fourth the volume of the mother cell. These cells are not only smaller, but they often differ from growing cells in other ways as well. In the case of *Tetrahymena* they have more cilia and a more slender shape; in choanoflagellates, these "swarmer cells" are devoid of the tentacular collar. In all cases these cells have a considerably higher motility than feeding cells (Fenchel, 1982b,c; Nelsen & DeBault, 1978; see Figure 4.3). This phenomenon is an ad-

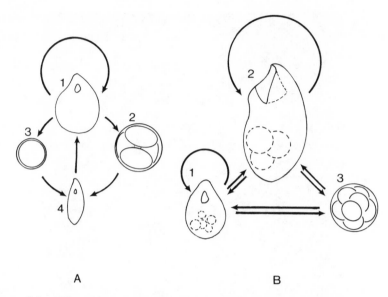

A                                    B

**Figure 5.2**   Life cycles of two species of *Tetrahymena*. A: *T. rostrata*. Trophonts (1) may divide to form new trophonts, they may form reproductive cysts
(2) or resting cysts (3). Excystment leads to tomites (4), which, if they find
food, develop into trophonts. B: *T. patula*. There are two types of trophic individuals, microstomes (1) and macrostomes (2); both may divide to reproduce
their own phenotype or they may transform into the other type. Both types also
form reproductive cysts (3). (Redrawn from Corliss, 1973.)

aptation to spatially patchy food resources. After a local food source is
depleted, the starving cell increases the probability that a copy of its
genome will reach another patch with food if it produces, for example,
four motile offspring at the onset of starvation. On the other hand, it is
probable that these smaller swarmer cells have a lower probability of
survival than a larger cell would have, but as long as their survival
probability is higher than 25 percent of that of the mother cell (in the
case of four swarmer cells being produced) this behavior should be
selectively favored.

A more advanced adaptation to spatially and temporally patchy environments is the evolution of a polymorphic life cycle, which includes
a trophic stage, called the "trophont," a (usually encysted) stage, in
which the ingested food is digested, and synthetic processes and several
successive cell divisions take place ("tomont") again leading to small,
rapidly swimming swarmer cells ("tomites") which search for a new
food patch. In some forms, such as *Tetrahymena rostrata* (Figure 5.2,A),
trophonts can reproduce by binary fission producing new trophonts,
and tomite formation only takes place if food is depleted. In other forms

(e.g., the soil ciliate *Colpoda* or some histophagous ciliates such as *Ophryoglena*) the mature trophont always forms a reproductive cyst in which multiple divisions take place. The adaptive significance of such life cycles is clear. The trophont can concentrate on feeding while food is there. However, the trophic stage is of short duration. In *Ophryoglena*, for example, the trophont feeds on a damaged worm for only about 30 minutes: During this time, however, it increases in cell volume by a factor of ten or more, until it is really only a film of cytoplasm covering a few huge feeding vacuoles. The resulting, encysted stage in which multiple cell divisions take place lasts for 12 to 24 hours. The duration of the tomite stage depends on when or if the cells find a new prey organism, but tomites can survive and remain viable for several days while searching for a wounded small metazoan (see Figure 3.9). This type of life cycle has developed independently within many ciliate groups; it seems to be a prerequisite for the evolution of parasitic life in some forms. Thus, a relative of *Ophryoglena* called *Ichthyophtirius* is a parasite of the skin of fishes, while several species of *Tetrahymena* are endoparasites of aquatic invertebrates (Batson, 1985; Canella & Rocchi-Canella, 1976; Corliss, 1973).

Another type of polymorphic life cycle can be considered as a mechanism for changing the food niche of an organism in response to the availability of food particles of diverse sizes. The predatory ciliate, *Didinium*, adjusts its cell size according to the size of the ciliate prey offered (Hewett, 1980). Ciliates belonging to the genus *Blepharisma* feed on bacteria and small flagellates. However, in batch cultures, after these food items have been depleted, giant blepharismas appear. In cultures these tend to be cannibalistic and this has led to anthropomorphic speculations that it represents some sort of voluntary "population regulation" (Giese, 1973). *Blepharisma* will, in fact, ingest anything of the right size and the high degree of cannibalism is only an artifact of dense batch cultures. The dimorphism can be seen as an adaptation to the microbial succession of decaying organic material. At first bacteria and small protozoa occur, but eventually larger organisms dominate, so by switching to giant forms, the blepharismas can utilize the patch for a longer time.

A similar phenomenon is the formation of "macrostome" cells. This occurs in otherwise bacterivorous species belonging to several genera of ciliates, but it is best studied in some species of *Tetrahymena* (Corliss, 1973; Figure 5.2,B). When "microstome" forms are depleted of their bacterial food, they develop macrostomes which can feed on other ciliates.

There is at least one example in which phenotypic polymorphism in a protozoan is a response to a predator rather than prey. In the hypotrich ciliate, *Euplotes*, individuals with a widely extended and keeled dorsal side are sometimes observed. Kuhlmann & Heckmann (1985) have recently shown that these morphological changes are induced by the presence of the predatory ciliate, *Lembadion*, and that they offer *Euplotes* some protection against being eaten.

Finally, polymorphic life cycles are found in some sessile protozoa which have motile "larval" forms. Suctoria, which are attached and devoid of cilia as adults, reproduce by a budding mechanism to form ciliated swarmer cells. When sessile, peritrich ciliates divide, one daughter cell develops a ciliary girdle. This "telotroch larva" swims off and attaches elsewhere to build a new stalk. Under adverse conditions mature, nondividing cells may also develop a ciliary girdle and leave the stalk to find more favorable conditions.

## 5.2    The adaptive significance of sexual processes

Among vascular plants and, in particular, among metazoa, sexuality is coupled to reproduction and is so widespread that many biologists rarely question its adaptive nature. However, suppose that in a sexual species a mutation occurs which causes females to reproduce by parthenogenesis. The female carrier of this gene would immediately enjoy an increase in Darwinian fitness by a factor of two, so parthenogenesis should quickly become established in the population. This is a somewhat simplified version of an argument which demonstrates that it is not a trivial question to ask why sexual processes and recombination occur in most higher organisms (Maynard Smith, 1978; Williams, 1975). It is generally held that there are a number of mechanisms which tend to maintain sexuality. These include the absence of genetic variance for nonsexual reproductive mechanisms; the coupling of meiosis with DNA repair; inbreeding depression in fitness in diploid organisms with selfing; and also some mechanisms of a more ecological nature. Thus, it is believed that recombination and sexual processes in a changing environment accelerate the fixation of more favorable mutant alleles. This latter mechanism should explain why asexual species or varieties among higher organisms usually have close, extant relatives capable of sexual reproduction. The conclusion is that asexual forms arise frequently among many groups of organisms, but on an evolutionary time scale they are short-lived since they respond more slowly to directional selection in a changing environment. Also a number of ecological situations have been considered in which the fitness of a sexual form producing a wide variety of genotypes is superior to an asexual competitor which produces a higher number of identical genotypes. This could occur, for instance, when the offspring compete for an environmental patch which has unpredictable environmental conditions.

Protozoa have historically played only a modest role in discussions on the adaptive nature of sex. This is regrettable because the distribution of sexuality among protozoa and their diversity of sexual processes show patterns which differ considerably from that found in higher eukaryotic organisms. This observation, as well as the fact that protozoa are well suited for an experimental approach, suggests that they could throw a new light on the significance of sexuality.

Sexuality probably arose among primitive eukaryotes. It may be speculated that it is based on two independent, evolutionary steps. One was the origin of recombination which may be understood as a DNA repair mechanism. The other, and probably later, step could be the evolution of heterokaryon forms, which could arise by cannibalism and later evolve into a diploid condition. It is conceivable that the heterokaryon could enjoy advantages over the haploid ancestor. It is not known whether sexuality arose only once (and was subsequently lost in some eukaryote lineages) or whether it evolved independently in different lineages. The variation in the processes of mitosis and meiosis among different groups of existing eukaryotes do not offer an unequivocal answer to this question. It is not the intention to give a general account of sexuality in protozoa here; for this the reader should consult general textbooks on protozoology, in particular Grell (1973) and, for cytological aspects, Raikov (1982). We will consider a few examples which throw some light on the role of sexuality in relation to the ecology of different species.

Sexuality is unevenly distributed among the major groups of protozoa, some of which do not seem to have this ability at all. The absence of sexuality is not easy to establish with certainty, since sexuality may be quite difficult to observe for a given group. Thus sexuality has only recently been discovered in some groups, such as dinoflagellates, where, in fact, it may be more widespread than was hitherto believed (Beam & Himes, 1980). Nevertheless, it seems certain that sexuality is absent in lobose amoebae and in many flagellate groups, such as the choanoflagellates, the kinetoplastids, the euglenids, and the chrysomonads. Among the sarcodines, obligatory selfing occurs in the diploid, actinophryid heliozoans. Sexuality probably occurs among the radiolaria, but it is not understood in detail. Among the foraminifera, alternating haploid and diploid generations are the rule; this life cycle implies gamete fusion, so that the mating system may be either one of selfing or of outbreeding, according to the species. Ciliates are diploid and sexuality in the form of conjugation or autogamy probably occurs in most forms. However, there are groups of ciliates, at least at the family level, in which sexuality has never been observed, and amicronucleate forms which must be asexual are known to occur in nature. The cytology of the sexual processes in ciliates is generally such that conjugation leads to two genotypically identical exconjugants, whereas autogamy leads to homozygosity at all loci. There are exceptions to this: In *Euplotes*, a mitotic division precedes meiosis, so that the two resulting generative, haploid nuclei within each conjugant need not be identical. Also, by a mechanism which is not understood, "autogamy" in *Euplotes* does not lead to any change in the genotype relative to the (heterozygous) parental cell (Dini, 1984), but the finding suggests that recombination does not take place. Due to the thorough analysis of mating systems and sexuality, particularly within species complexes belonging to *Paramecium, Tetrahymena*, and *Euplotes*, some considerations of the role of sexuality can be offered in the case of ciliates.

Among these genera, obligatory outbreeders—forms with obligatory or facultative autogamy—as well as asexual forms occur. In the *Paramecium aurelia* complex, most sibling species are obligatory outbreeders (Sonneborn, 1957, 1975). Clonal cultures do not undergo sexual processes, since these processes require different mating types (in this case there are only two mating types per species, but in other forms a large number of mating types may exist). Furthermore, some of the species have periods of "immaturity" (lasting for a certain number of generations) during which they cannot conjugate. These traits suggest that there is selection for outbreeding; the period of immaturity increases the probability that the cells will move around for some time between conjugations, and thus inbreeding becomes less probable. However, obligatory autogamy also occurs in some strains in which all individuals are homozygous at all loci.

It is likely that autogamy in ciliates originated in small, isolated populations in which a high degree of inbreeding had already occurred. After some time, total loss of outbreeding would not affect fitness and autogamy became established. In a diploid organism, inbreeding leads to an initial depression in fitness since recessive, deleterious genes can affect the phenotype in the homozygous condition. However, if the inbreeding strain survives, its fitness will eventually increase again since selection against deleterious genes is more efficient in homozygotes. At this stage autogamy may become established without any loss in fitness.

Among the species of the *Paramecium aurelia* complex (and in other ciliates with sexuality that have been studied) a phenomenon referred to as "clonal senility" is found. If a clonal culture with sexually competent cells is maintained for a long time, cell viability eventually deteriorates. This can be counteracted only by conjugation with cells belonging to another mating type or by autogamy (if it occurs in the strain in question). This suggests that DNA repair is coupled to the process of meiosis of the micronucleus and that periodic renewal of the macronucleus is necessary; it also explains why autogamy, rather than an entirely sexless condition, is maintained in some of the sibling species.

The entirely asexual ciliates, including the amicronucleate strains of *Tetrahymena* (which according to Nanney, 1982, must have an independent existence over a considerable time span), have apparently transferred the ability to repair DNA to the macronucleus, thus making themselves independent of the process of meiosis.

In *Euplotes vannus*, obligatory outbreeding strains occur, as well as strains which normally undergo autogamy (Dini, 1984; Nobili et al., 1978). The outbreeding type is comprised of a high number of mating types which are controlled by a multiple allele system: These ciliates have a long period of immaturity suggesting selection against inbreeding. The strains which can perform autogamy are also capable of conjugation with outbreeding cells. It has thus been established that the ability to perform autogamy is controlled by a single allele. The outbreeding form shows a widespread and continuous distribution in marine environments. In contrast, the autogamous ciliates are rarer and form small, isolated populations. This observation suggests that, in general, outbreeding in large populations confers some selective advantage over a short time scale (in evolutionary terms). It also supports the above-mentioned model for the origin of autogamy in small, isolated populations.

There are a few observations which suggest that recombination and other sexual processes do confer some selective value under changing environmental conditions. In free-living ciliates, conjugation is often induced in populations which are exposed to adverse conditions. In nonclonal batch cultures of many species, an epidemic of conjugation often occurs following the onset of starvation and the end of the exponential growth phase (Bick, 1966; Nobili et al., 1978). This is reminiscent of the situation for many arthropods and rotifers in which sexuality occurs only at the end of the summer season following many generations of parthenogenetic reproduction. To what extent the intuitive interpretation (namely, that a reshuffling of the genotypes will confer increased fitness to the original cell clones when exposed to adverse or unpredictable conditions) will stand a more critical analysis is not known.

There have been two experimental attempts to study the effects on niche width and tolerance to environmental factors in ciliates of different mating systems (Dini, 1981; Nyberg, 1974). In both cases, strains with various degrees of outbreeding and inbreeding of different species were exposed to various levels of a variety of ions of heavy metals, to decreased pH, and to increased temperature. In all cases the outbreeders seemed to have a wider range of tolerance to the experimental conditions tested. The interpretation of these experiments is not straightforward, and could, in general, not be the result of heterosis. These experiments illustrate the wide and largely unexploited scope of studies on the significance of sexuality in protozoa.

## 5.3    The physical and chemical environment

Studies of tolerance to various environmental factors has sometimes been considered to be almost synonymous with ecology. The topic has been studied by correlating distribution patterns with environmental factors, such as salinity, pH, and temperature. The findings may or may not reflect causal relationships. Alternatively, the survival or growth of protozoa as a function of some chemical or physical factor has been extensively studied in the laboratory. A vast amount of data have accumulated in the literature, much of which is reviewed in Noland & Gojdics (1967). The topic will be treated briefly here because many aspects are discussed more effectively in other contexts, in particular, when discussing the habitats of protozoa, and also this approach yields only limited ecological and physiological insight.

Water is an absolute requirement for certain protozoan life. Within the aquatic environment, certain ranges of different physical and chemical factors are required for protozoan life. Among such factors we will look primarily at temperature and at electrolyte and hydrogen ion concentration.

Much evidence shows that clonal cultures can tolerate a certain increase or decrease in temperature and slow acclimatization extends this range. However, there is a genotypically fixed range for isolates, beyond which growth and survival does not take place, and also a genotypically fixed optimum temperature at which growth rate takes a maximum value. This optimum temperature is close to the maximum temperature which the strain can endure. Over a fairly wide span of temperatures below the optimum temperature the $Q_{10}$ (the factor by which rate constants are increased when temperature is increased by 10°C) is within the range of 2 to 3; at lower temperatures, $Q_{10}$ increases strongly (Figure 5.3).

The genotypically fixed temperature range of growth and survival and the optimum temperature both correlate with the natural habitats of the organisms. Thus, protozoa isolated from tropical habitats show a temperature range for growth which is higher than that found in forms isolated for temperate habitats, while some ciliates isolated from seawater ice from the Antarctic have much lower temperature optima. Among freshwater protozoa found in the temperate zone, there is some correlation between their response to temperature in the laboratory and their seasonal occurrence in lakes (Dragesco, 1968; C.C. Lee & Fenchel, 1972; Noland & Gojdics, 1967; Figure 5.3).

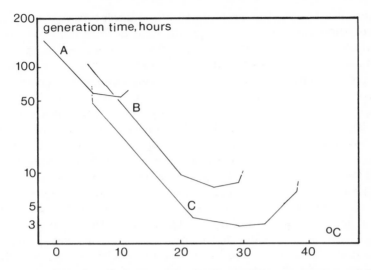

**Figure 5.3**   Minimum generation time as a function of temperature in clonal cultures of the hypotrich ciliate *Euplotes* isolated from Antarctic sea ice (A), from Denmark (B), and from Florida (C). (Redrawn from Lee & Fenchel, 1972.)

Some protozoa are active at sub-zero temperatures. At the other extreme, such as in hot springs, protozoa are found up to about 50°C. Issel (1910) recorded protozoa from thermal springs in Italy and found that the amoeba *Hyalodiscus* and the ciliate *Cyclidium* could be found at temperatures exceeding 50°C, but most forms occurred only at lower temperatures. Similar results were obtained by Nisbet (1984) in thermal springs in New Zealand. Some protozoa isolated from shallow ponds in temperate climates can also be adapted to grow at temperatures of around 50°C (Dingfelder, 1962).

Protozoa regulate their electrolyte composition and osmotic pressure by three mechanisms: active ion transport across the cell membrane; regulation of the concentration of low-molecular weight solutes, mainly amino acids, in the cytoplasm; and by active excretion of water. This latter process probably takes place in the "spongiome," a system of endoplasmic membranes, and the water is then in most cases collected in a contractile vacuole before being expelled from the cell. The physiology of osmoregulation and contractile vacuoles was reviewed by Patterson (1980). Freshwater protozoa are hyperosmotic relative to their surroundings, so they must counteract swelling by the active removal of water from cells, for which purpose they nearly always have a contractile vacuole. Marine ciliates usually also possess a contractile vacuole, but many other marine protozoa do not.

There is a remarkable variation among different protozoan groups as regards their tolerance to changes in salinity. The radiolaria and the acantharia are stenohaline organisms virtually confined to oceanic environments; this is also the case of planktonic foraminifera. Benthic, calcareous foraminifera are also, in general, stenohaline; however, several species penetrate into brackish waters (Anderson, 1983; Lutze, 1965; Rottgardt, 1952; Zeitzschel, 1982). In contrast, many ciliates, lobose amoebae, and both marine and freshwater flagellates, are amazingly euryhaline. There are many reports showing that a number of freshwater ciliates can adapt to live in seawater, or at least diluted seawater, and many marine ciliates can tolerate very dilute, brackish water (Ax & Ax, 1960; Bick, 1964; Finley, 1930; Lackey, 1938a). In fact, there are probably a few species of flagellates and ciliates which occur in freshwater as well as in seawater. In dilute, brackish waters the usual ciliate fauna is a mixture of freshwater and marine species. However, at least one major group of ciliates, the karyorelictids, includes only stenohaline forms; except for the limnic genus, *Loxodes*, they are all strictly marine and do not occur in brackish water. This may be related to the fact that the karyorelictids do not have a contractile vacuole (Fenchel, 1969). The great majority of thecamoebans are confined to freshwater, but marine species do exist. Among the heliozoa, both marine and limnic species occur, but there is no experimental evidence on the adaptability of these forms to changes in osmotic pressure of the environment.

There seem to be very few studies of the protozoan fauna of either hyperhaline or alkaline lakes. Kirby (1934) and Pack (1919) described ciliate faunas from hyperhaline environments with salinities exceeding 10 percent. Finally, Curds et al. (1986) found a rich ciliate fauna in some African soda lakes, containing $Na^+$ and $HCO_3^- \text{-} CO_3^{2-}$ as the principle ions, at a pH of around 10. Protozoan tolerance to extremely low pH values seems to be poorly documented in the literature. Lackey (1938b) observed several species in acid mine drainages at pH values within the range of 2 to 3. The green flagellate *Euglena mutabilis* is especially characteristic for acid waters. It should be borne in mind, however, that significant changes in pH are always associated with changes in a plethora of other chemical factors.

In environments with high temperatures, as well as in hyperhaline or acid waters, species richness is greatly reduced. This is often taken as an evidence of "extreme" environmental conditions *per se*. The fact that protozoa are found in such environments demonstrates that eukaryotic organisms can adapt to these conditions. The low number of species found in such environments is probably a consequence of the small size of such habitats, as well as their ephemeral nature, on an evolutionary time scale. In comparison with prokaryotes, however, eukaryotic organisms are more limited by fundamental, physiological constraints

with regard to tolerance of the most extreme environments. Some prokaryotes can live at temperatures exceeding 90 °C; others (*Thiobacillus*)
can grow at values of pH below 1; while halophilic bacteria can thrive
in concentrated brine (Brock, 1969, 1978). Thus, a few aquatic environments exist which do not harbor protozoa (or other eukaryotes),
although prokaryotes are found in them.

# 6

# Symbiosis

## 6.1 The definition of symbiosis

The term "symbiosis" does not have the same meaning to all biologists. Here we use it in the literal sense of "living together," that is, to describe any physically close association between different species irrespective of the functional significance. Such a general term is necessary because there are many cases where the functional nature of the relationship is not yet understood. Symbiosis thus includes "parasitism," in which one member of the association, the host, suffers while the parasite gains from the association; "mutualism" in which both components gain from the association; and "commensalism" in which the association is neutral from the viewpoint of one of the components. In the real world, there are gradients between these classes of symbiotic associations; the degree of benefit or harm involved may vary, according to environmental conditions or the genotypes of the components.

It is customary to distinguish between endo—and ectosymbionts; the former live inside the host cell while the latter are found on the cell surface. Again, this classification is not always clear cut; endosymbionts are mostly, but not always, found inside vacuoles lined with host-cell membranes, while in other cases, host cells contain recesses in their external cell membrane or mucous surface layer which harbor ectosymbionts.

A vast number of associations between different types of protists, between protists and prokaryotes, and between protozoa and metazoa are described in the literature; among these, relatively few have been studied experimentally. The evolution of protozoan associations with metazoa will be discussed in Chapter 11. Here we will consider only those associations in which protozoa harbor other protists or prokaryotes.

During the last decade, endosymbionts of protozoa have drawn a considerable amount of attention. This reflects, in part, the interest in theories on the origin of eukaryotic cells through endosymbiotic associations and, in part, the recognition of the ecological significance of symbiosis. The subject has even been considered as a separate, biological discipline for which the term "cytobiology" has been coined (see Lee & Corliss, 1985, for references).

## 6.2   Associations with photosynthetic organisms

The presence of photosynthetic symbionts in cells of protozoa is a widespread phenomenon which, in some cases, has a considerable impact on the function of ecosystems. It is customary to classify the symbionts as either "cyanella," which are cyanobacteria; "zoochlorella" (or green symbionts), which include chlorophytes, prasinomonads, and volvocids; and "zooxanthella" (or yellow symbionts), which include dinoflagellates, diatoms, chrysomonads, and prymnesids. The relatively few examples of cryptomonad endosymbionts do not fit into this classification. It is not always a simple matter to establish the taxonomic nature of an endosymbiont. The composition of their photosynthetic pigments yields some information, but the morphology of the symbiont may be considerably altered as compared to its free-living relatives. Endosymbionts often are found in cyst-like stages without flagella; endosymbiotic diatoms may not form frustules. However, occasionally swarmer stages may occur, or it may be possible to grow the symbionts outside of the host, in which case their taxonomic affinity can be determined. Some symbiotic associations are obligatory for both members, neither of which can be grown on its own; in these cases, the morphology of the symbiont has often evolved into one which is radically different from its free-living relatives.

The functional significance of symbiosis between a photosynthetic and phagotrophic organism seems intuitively obvious. The phagotrophic heterotrophic component obtains reduced carbon either in the form of carbohydrate products of photosynthesis or by digesting the symbionts. The photosynthetic component, on the other hand, receives mineral nutrients and increased motility from the host (the motile behavior of which is often modified to favor the symbiont). In protozoa with calcareous shells, the calcification process may be facilitated by photosynthetic symbionts. This is because the bicarbonate-carbonate equilibrium is affected by photosynthesis. All these mechanisms have been established experimentally in many cases.

This sort of symbiotic relationship must have originated as a result of the phagocytosis of photosynthetic cells, which in some way were able to evade digestion and to become established in a vacuole in the host cytoplasm. In some cases, the co-evolution of the host and its endosymbiont has led to a substantial modification of both components, so that the symbionts are transferred to the next host generation through a form of cytoplasmic inheritance. There are, however, many cases in which the endosymbionts also occur as free-living forms, or as swarmers, and new host cells are infected by ingestion of free-living symbiont cells. In foraminifera and radiolaria, the symbionts are not transferred with the gametes, so that every generation arising from a zygote cell must be infected anew. There are also examples of apochlorotic flagellates, which secondarily have acquired photosynthetic capabilities as a result of symbiosis with a photosynthetic organism. The dinoflagellate *Peridinium balticum* was thought to have two different types of nuclei. Closer analysis showed that the chloroplast is enclosed, together with one of the nuclei, in a single membrane. The flagellate is thus a compound organism consisting of a dinoflagellate and an endosymbiotic (euglenid) flagellate. There are vestiges of a dinoflagellate chloroplast as well, suggesting that the dinoflagellate ancestor once had the capability of photosynthesis on its own (Taylor, 1982).

Cyanella (that is, endosymbiotic cyanobacteria) are rarely found among protozoa. One established and closely studied example represents the intermediate situation between a symbiotic relationship of a eukaryote and a cyanobacterium on the one hand and a flagellate with chloroplasts on the other. *Cyanophora paradoxum* is a flagellate of obscure taxonomic position, but it is often referred to as a cryptomonad. It harbors two cyanella; the association is obligatory for both components and highly integrated, but in many respects the cyanella are simply cyanobacteria which are devoid of a cell wall. Their association has been studied in detail by Trench et al. (1978). The great majority of endosymbiotic photosynthetic organisms in protozoa is comprised of eukaryotic flagellates or algae, which will be discussed below.

It could be argued that symbiosis between phagotrophic and photosynthetic organisms would be especially favored in oligotrophic waters, where a closed nutrient cycle within the symbiotic consortium would confer advantages on both parts. This is supported by the fact that all large planktonic protozoa in the euphotic zone of the oceans seem to harbor endosymbiotic algae or flagellates. Thus, two of the main groups of radiolaria (the spumellaria and the nassellaria), as well as the acantharia and the planktonic foraminifera, all have photosynthetic symbionts. (The phaeodarian radiolaria are mainly deep-water forms and do not harbor symbionts.) The symbionts of most radiolaria are dinoflagellates, but some spumellaria harbor prasinomonads instead (Anderson,

1983; Cachon & Caran, 1979). Symbionts occur in great numbers in the ectoplasm and show diurnal migration within the host cell; in light they move outward, while in dark they move toward the center of the host. Similar migrations occur in acantharia and foraminifera. The acantharia harbor prymnesids as symbionts (Febvre & Febvre-Chevalier, 1979). Most planktonic foraminifera harbor dinoflagellates, but some forms carry prymnesids instead. Each foraminiferan cell contains many hundreds of symbionts in the peripheral pseudopodial network. In addition, some planktonic foraminifera have ectosymbiotic dinoflagellates (Anderson & Bé, 1976b; F.J.R. Taylor, 1982).

All these planktonic sarcodines are primarily phagotrophs. However, in the case of radiolaria, the transfer of $^{14}$C-labelled photosynthates to the host cytoplasm has been demonstrated and there is evidence that they also digest symbiont cells. In the laboratory, survival is enhanced in the light and there is little doubt that the symbionts are of nutritive significance to the host. Measurements of the photosynthetic rates of the protozoa show, that in terms of the level of chlorophyll $a$, they are comparable to or exceed those of most phytoplankton cells (Anderson, 1983; F.J.R. Taylor, 1982, and references therein).

The giant benthic foraminifera belonging to the families Nummulitidae, Amphisteginidae, and Soritidae (Figure 6.1,B) all harbor photosynthetic symbionts; in the first two groups these are diatoms, while among the soritids, dinoflagellates, chlorophytes, or rhodophytes occur. The giant foraminifera are all inhabitants of shallow, tropical seas. Like reef-building corals, they are of biogeochemical significance for the formation of carbonate sediments. Evidence for this comes not only from their extreme abundance in reef flats, seagrass meadows, and other shallow-water habitats, but also from the existence of fossil limestone deposits based on foraminiferan shells. Thus the Cheops Pyramid is built from the shells of the giant Eocene foraminifer, *Nummulites*. It is generally believed that the evolution of giant foraminifera required their symbiosis with photosynthetic algae which assisted in calcification and serve as an internal $O_2$ supply (see Section 1.6). (Giant foraminifera devoid of symbionts and belonging to other groups occur in sediments at greater depth, and also in temperate waters, but they do not make calcareous shells.) The biology of large foraminifera has been studied in considerable detail as was recently reviewed by Lee (1983) and Lee & McEnery (1983), on whose reviews the following account is based.

As in the case of planktonic sarcodines, symbiosis is re-established in each generation. Specificity between host and symbiont species seems to be surprisingly low, although within local species populations the majority of individuals harbor the same species of symbionts. Endosymbiotic diatoms of foraminifera seem to belong to otherwise free-living species. In the host cytoplasm they lose their frustule, but if isolated

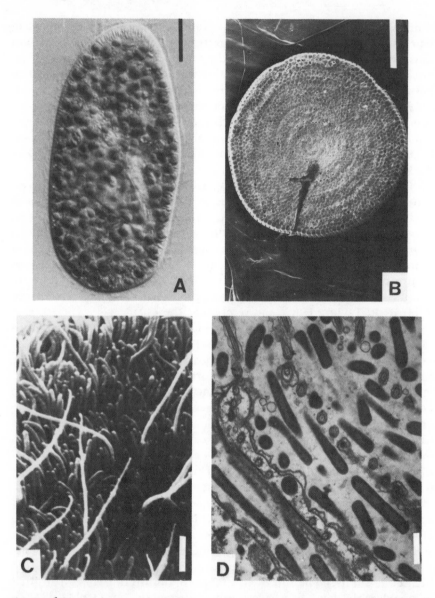

**Figure 6.1**  Symbionts of protozoa. A *Paramecium bursaria* with *Chlorella* cells. B. The giant foraminifer *Sorites* sp. (collected from the Red Sea, Eilath, Israel). C,D. Ectosymbiotic bacteria on the surface of anaerobic ciliates. C. The surface of *Parablepharisma pellitum* with bacteria oriented perpendicularly to the cell surface and with cilia protruding through the bacterial layer (scanning-electron micrograph). D. Transmission-electron micrograph of the surface of a *Sonderia* sp., showing bacteria lying parallel to the cell surface between the ciliary rows. Scale bars: A: 20 μm, B: 1 mm, C: 2 μm, D: 1 μm. (C,D, after Fenchel et al., 1977.)

from the host and brought into culture the diatoms regain the same appearance as the free-living forms. The foraminiferan shells show a number of morphological adaptations which seem to favor the symbionts. These include thin, translucent parts of the shell under which the symbionts are situated. Much evidence demonstrates the nutritive significance of the symbionts for the host cells; most convincing perhaps, is the fact that at least some of these foraminiferan species can be grown with light as their only source of energy and that growth is proportional to illumination within a certain range of intensities (Röttger, 1972; Röttger & Berger, 1972). Direct transfer of photosynthate from the symbionts to the host has also been demonstrated with radio-tracer methods (Kremer et al., 1980; Muller, 1978) and the significance of photosynthesis for the rate of calcification has also been demonstrated experimentally (Erez, 1978).

In smaller foraminifera belonging to the genera *Elphidium* (see Figure 3.10) and *Nonion*, another phenomenon, referred to as "chloroplast-symbiosis," is found. The name is not well chosen because this process is not true symbiosis. These foraminifera feed on microalgae on the sediment surface, in particular, diatoms. However, they do not digest the chloroplasts of the prey: Instead they are retained in a functional state in vacuoles, sometimes for several weeks. Incubation of the foraminifera in the light with $^{14}$C-labelled bicarbonate demonstrated that the prey chloroplasts remain functional and contribute significantly to the nutrition of the foraminifera (Lopez, 1979).

Symbiosis with photosynthetic organisms is found in many ciliate species, but its occurrence does not seem to correlate closely with taxonomic position of the host or with special types of habitats. The classical example is *Paramecium bursaria* (Figure 6.1,A) which has been studied experimentally in some detail. In nature the ciliates nearly always harbor hundreds of *Chlorella* cells situated in special "perialgal" vacuoles. In the laboratory, aposymbiotic individuals can be produced by growing the ciliates in the dark with bacteria as food. Under these circumstances the algal cells cannot multiply and are outgrown by the ciliates and eventually lost after a sufficient number of host generations. The symbiotic algae can also be grown outside the host. Reinfection takes place by phagocytosis of suspended *Chlorella* cells. Due to some recognition mechanism which is not understood, food vacuoles containing algae do not fuse with lysosomes. Thus the algae evade digestion and the food vacuoles become perialgal vacuoles. The ciliate may be infected with various strains of *Chlorella*, including strains which have been isolated directly from lake water. The various strains, however, differ with respect to the success with which they colonize the host. In the case of normally free-living strains, a large number of cells will be digested following phagocytosis, while those which do become established as symbionts are lost more rapidly during periods of darkness, than is the case for symbiotic strains. Paramecia with symbionts are capable of growing in an appropriate mineral solution in the absence of food if exposed to sufficient illumination. The organic materials transferred from the symbionts to the host mainly include carbohydrates, in particular maltose. Using mechanisms which are not yet understood, under normal conditions the ciliate can maintain a constant number of symbionts irrespective of growth rate (Brown & Nielsen, 1974; Karakashian, 1975; Muscatine & Poole, 1979, and references therein).

There are several other examples, mainly among freshwater ciliates, of similar associations and *Chlorella* seems to be the most frequent symbiont. These ciliates include species of *Stentor, Frontonia, Strombidium, Euplotes* (Figure 9.2,A), and *Climacostomum*. One species of the marine genus *Condylostoma* harbors cells of the volvocid, *Chlamydomonas*. Christopher & Patterson (1983) listed freshwater protozoa which have been reported as harboring zoochlorella. The list includes 42 species of ciliates, 21 species of testate amoebae, 5 species of heliozoa, and 4 species of lobose amoebae.

By far the most remarkable case of photosynthetic endosymbiosis in ciliates occurs in *Mesodinium rubrum*. This ciliate is a frequent inhabitant of the plankton of all seas. Locally it may occur at extremely high densities (up to $10^5$ cells per ml), producing "red tides." This phenomenon has been recorded from most coastal seas. From their writings it seems that both Leeuwenhoek (in 1676 off the coast of Holland), and later, Darwin (in 1839 on the Beagle off the coast of South America) observed blooms of *Mesodinium* (Taylor et al., 1971).

*Mesodinium rubrum* cells are always packed with chloroplasts, the pigments and structure of which suggest a cryptomonad affinity. The actual structure of the symbiont within the ciliate has been controversial and is, perhaps, still not quite clarified. Taylor et al. (1971), on the basis of transmission-electron microscopy, found no symbiont nucleus, but only isolated chloroplasts surrounded by some cytoplasm, a few mitochondria, and a cell membrane. This observation, of course, is difficult to understand, unless the ciliate acquires new "symbionts" at regular intervals (namely, as an example of "chloroplast symbiosis"), but it probably does not ingest particulate material at all. Hibberd (1977) found that there is actually one symbiont nucleus per ciliate, but that the symbiont is branched in a complicated manner, so that using only a limited number of sections for microscopy it is easy to arrive at the conclusion of Taylor et al. (op.cit.). More recently, Oakley & Taylor (1978) found that while there is a nucleus, individual branches of the symbiont containing only a chloroplast and some cytoplasm may become disconnected from the nucleus-containing part of the symbiont. Both the structure of the mouth and the absence of feeding vacuoles suggest that *Mesodinium rubrum* depends entirely on the endosymbiont and that it has, in fact, become a photosynthetic ciliate. As such, it plays a considerable role as a primary producer under some circumstances (Packard et al., 1978).

There is also evidence that certain marine ciliates retain the chloroplasts of their prey for some time in a functional state as do the benthic foraminifera discussed above (Blackbourn et al., 1973). However, there is no experimental evidence regarding the functional significance of these observations.

## 6.3   Nonphotosynthetic symbionts

Like other eukaryotes, protozoa may themselves fall victim to protozoan parasites. For example, some suctorians are endoparasites in other ciliates (e.g., Jankowski, 1963), trypanosomatid flagellates may invade the nucleus of ciliates (Wille et al., 1981) and another type of flagellate may occur as ectoparasites on ciliates (Foissner & Foissner, 1984). Associations between heterotrophic prokaryotes and protozoa are more common. Light- and electron-microscopic observations reveal that many or perhaps most protozoa carry endo- and/or ectosymbiotic bacterial cells. In most cases nothing is known about their functional significance. In a few cases, it has been found that protozoa with endosymbiotic bacteria do not tolerate antibiotics and this has been interpreted as an indication of the vital importance of the prokaryote for the host. A remarkable system was studied by Jeon (1983, and papers cited therein). He studied a laboratory strain of an amoeba which accidentally became infected by bacteria. The infection proved lethal to most amoebae, but the survivors maintained bacteria in their cytoplasm. After several years the association had become obligatory.

In some species of ciliates, such as within the genus *Paramecium*, the biology of endosymbiotic bacteria has been studied in more detail. In *Paramecium* there are several types of such bacteria which reside either in the nucleus or in the cytoplasm. Many of their properties were known before they were recognized to be bacteria so they were originally designated by letters from the Greek alphabet (e.g., "Kappa particles") and studied as an example of cytoplasmic inheritance (Sonneborn, 1959). Subsequently they were shown to be Gram-negative bacteria (Preer et al., 1974). Best known are the Kappa particles which are now assigned to the genus *Caedibacter*. Their survival in the cytoplasm of paramecia requires the presence of a dominant protozoal allele, K, so that only paramecia of the genotypes KK or Kk can be carriers of these bacteria. Such carriers are referred to as "killers" because the bacteria produce "refractile bodies" which are released into the surrounding water. These bodies are toxic and if "sensitive" paramecia ingest them, the ciliates are killed. Clearly, this symbiotic relationship confers an advantage to the carriers as long as sensitive cells are present, but it is unclear why the k gene is maintained in populations of *Paramecium*. Other similar systems have been described for paramecia and for other ciliates. A peculiar one is the "mate-killing trait" in which the carrier kills sensitive mates during conjugation (Dini & Luporini, 1982). Further references to such systems can be found in Lee & Corliss (1985).

Although many protozoa contain symbiotic bacteria, these symbionts are especially common in protozoa from anaerobic environments. Not only do virtually all such protozoa harbor bacteria, but the number of bacteria per cell is very high. Endo- and ectosymbiotic bacteria of termite flagellates and ectosymbiotic bacteria on rumen ciliates have long been observed. Free-living ciliates from anaerobic environments also carry ecto- as well as endosymbiotic bacteria (Fenchel et al., 1977; see also Figure 6.1). These bacteria are of diverse morphological types and are sometimes pigmented; on the basis of microscopic observations each species of ciliate is associated with a unique species of bacteria. The ectosymbiotic bacteria may be aligned either perpendicularly to the cell surface, between the ciliary rows, and often inserted in depressions in the cell membrane; or they may be oriented parallel to the surface in a mucous layer formed by the host. The endosymbiotic bacteria on the other hand are distributed throughout the cytoplasm but are most abundant beneath the cell membrane. The total volume of these bacteria is considerable in comparison to the host volume. For example, one individual of a *Sonderia* species carries some $10^5$ ectosymbionts: these bacteria thus constitute about 20 percent of the total volume of the protozoan; in addition the number of endosymbionts is comparable to the number of bacteria which are found on the external surface of the ciliate.

It is likely that such masses of symbiotic bacteria must somehow affect the energy metabolism of their host. Since anaerobic protozoa generate hydrogen in their fermentative metabolism (Section 4.4), one possibility would be that the bacteria are methanogens which oxidize $H_2$ with $CO_2$. This would be of benefit to the ciliate in the same way that methanogenesis is necessary for maintaining the rate of fermentation in the cow rumen (Hungate, 1975). Methanogenic bacteria are characterized by fluorescent coenzymes with absorption peaks at 350 and 420 nm, and they can be identified by fluorescence microscopy using UV light of appropriate wave lengths. Thus, Vogels et al. (1980) showed that the ectosymbiont bacteria of rumen ciliates are methanogens. It was subsequently shown that some free-living, anaerobic ciliates (e.g.,

*Metopus* and *Caenomorpha*) contain fluorescent endosymbionts which presumably also are methanogenic bacteria (Van Bruggen et al., 1983).

While symbiosis obviously is favorable for bacteria living in an environment in which there is competition for hydrogen (for methanogenesis and for sulfate reduction) and while the fermenting ciliates gain from a low $H_2$ pressure (see Section 4.4), it is not yet known whether the protozoa also utilize the bacteria in other ways, such as by digesting them or by receiving dissolved organic materials from them. It is also not known whether, in fact, all bacterial symbionts of anaerobic protozoa are methanogens. Certain sediment-dwelling invertebrates, notably some pogonophorans and bivalves, contain chemolithotrophic sulfur bacteria which oxidize $HS^-$ to $S^0$ or to $SO_4^{2-}$ (Dando et al., 1985). Among ciliates a possible candidate for this process are species in the unusual ciliate genus *Kentrophorus*. These are entirely covered on one side with a dense layer of pigmented bacteria; these ciliates apparently do not form food vacuoles. Fauré-Fremiet (1950a) considered the bacteria to be "sulfur-bacteria," and the ciliate is found at the top of the anaerobic zone in sandy sediments (Fenchel, 1969).

A special case of bacterial endosymbiosis is that of *Pelomyxa palustris*. This is a giant, monopodial amoeba which occurs in the ooze of some ponds and lakes (Figure 2.3,B,C). Among various pecularities of this creature is the absence of mitochondria. (This and other features have inspired some authors to consider *Pelomyxa* to be a uniquely primitive eukaryote, see e.g., Corliss, 1984. However, these traits may just as well have evolved secondarily.) The amoeba does harbor three different kinds of endosymbiotic bacteria (Andresen et al., 1968). Recently, Van Bruggen et al. (1983) found that two of the bacterial types show fluorescence which would indicate that they are methanogens. This finding is somewhat at variance with the finding by Chapman-Andresen & Hamburger (1981) that *Pelomyxa* is an aerobic organism which respires oxygen (although the amoeba also occurs in environments with little or no oxygen) since methanogens are considered to be extremely sensitive to oxygen. The exact nature of the metabolism and of the symbiotic bacteria of *Pelomyxa* remains to be fully understood; unfortunately it has not yet been possible to grow this somewhat elusive amoeba under controlled conditions in the laboratory.

# 7

# The Niches of Protozoa

## 7.1 Introduction

This chapter contains some comprehensive ideas and principles about the role of protozoa in nature. These considerations are often discussed separately for different types of habitats, such as marine plankton or soils. In many respects the functional roles (niches) of protozoa do not differ fundamentally among different ecosystems and so it seems practical to discuss them within the common framework of theoretical ecology as an introduction to protozoan communities in nature.

The special ecological properties of protozoa derive primarily from their sizes or from factors which correlate with size. Obviously there is a definite correlation between the size of an organism and the size of its food particles, so protozoa play a key role in food chains based on bacteria or on the smallest photosynthetic eukaryotes. Small organisms have potential for rapid growth and a high metabolic rate; a relatively small protozoan biomass may therefore have a relatively large effect on element cycling. Natural ecosystems are uneven in time and space at all scales. In such heterogeneous environments, body sizes and potential growth rates become niche parameters. Due to their growth potential, protozoan populations can utilize rapid fluctuations in resource levels and due to their small size, protozoa can exploit very small patches of resources or tiny habitats. As shown in Figure 1.4, protozoa span a size range of about three orders of magnitude. In many habitats protozoa are therefore found at several trophic levels of food chains, while environmental heterogeneity from the viewpoint of a microflagellate is very different from that of a large sarcodine.

## 7.2 Steady-state phagotrophic food chains

While exceptions occur, for example, the histophagous ciliates previously discussed, predators are usually larger than their prey. There is also a correlation between the minimum size of prey particles and predator size. This was discussed in Section 3.1 for the case of suspended food particles and is illustrated in Figure 3.2. This figure shows that the typical length ratio between predator and prey is around ten: for suspension-feeding organisms there are probably very few cases in which the optimum food-particle size deviates from this ratio by more than a factor of about ten. The ciliate *Colpoda* which is about 100 times longer than the bacteria on which it feeds, is analogous to a ten-meter-long baleen whale catching ten-centimeter-long krill. To be sure, some aquatic invertebrates, such as *Daphnia* (Peterson et al., 1978) and larvacean tunicates (King et al., 1980), exploit large, suspended bacteria, but such examples seem to be exceptions (Fenchel, 1984). Attached bacteria are exploited by many small invertebrates while large invertebrates feeding on detritus or sediment utilize microorganisms ingested in bulk together with other material. Nevertheless, for the present purpose it is a valid generalization to consider biological communities as an assemblage of organisms in which the larger ones eat the smaller ones and in which there is a more or less fixed size ratio between predator and prey.

For such an idealized food chain in a steady state (which hardly ever occurs in nature), what is the ratio between the biomasses represented by the different trophic levels? This has been considered theoretically by Kerr (1974) and by Platt & Denman (1977). The result depends on the weight ratio between organisms of trophic level i and i + 1; the gross growth efficiency; and the parameters of the equations, $\mu = a_\mu W^b$ and $R = a_R W^b$, which describe the growth rate constant ($\mu$) and the weight-specific metabolic rate (R) as functions of body weight (W), respectively (see Section 4.2). For these equations, $a_\mu/(a_\mu + a_R)$ = net growth efficiency, and we assume that b = $-0.25$ (see Section 4.2). Using these assumptions and reasoning related to that of Kerr (op. cit.), it can be shown that the ratio between the total biomass, represented by trophic level i + 1 and i, is given by $(W_i/W_{i+1})^b$ E, where E is the gross growth efficiency. For reasonable estimates of the parameters (e.g., E = 0.3 and $W_i/W_{i+1}$ = 0.001) we find that the various trophic levels in a community should have roughly equal biomasses and that the number of individuals at each trophic level should be inversely proportional to their body sizes. This general result is not very sensitive to the size ratio between predator and prey within biologically reasonable limits. The gross growth efficiency is, of course, subject to some variation, but it is likely to be within the range of 0.3 and 0.5. Net growth efficiency is fixed at about 0.6. At each trophic level, therefore, approximately 30 percent of the ingested organic carbon is mineralized and approximately another 30 percent is egested or excreted as particulate or dissolved organic material, respectively; the remainder is incorporated into cells. In such an idealized microbial food chain with three trophic levels, for example, bacteria being consumed by flagellates which then are eaten by ciliates, about 16 percent of the bacterial production will end up as part of the ciliated cells; about 42 percent will become mineralized; and another 42 percent will be excreted or egested as dead organic material. The reason that the three trophic levels will represent approximately the same total biomass at any given moment is the result of the fact that the turnover rate of cell carbon is lower in larger organisms.

Platt & Denman's (1977) somewhat more sophisticated model, which considers the transfer of biomass along a continuum of particle sizes, yields similar results. The best evidence for the general validity of this type of model is provided by the data of Sheldon et al. (1972) which show that if all the living particles in oceanic pelagic communities (from bacteria to whales) are grouped into logarithmic size classes, each of the classes comprises about the same total biomass. Examples which will be discussed in Chapters 8, 9, and 10 show that in phagotrophic microbial food chains the different trophic levels represent comparable amounts of total biomass.

The role of microphagotrophic organisms in the remineralization of organic nitrogen and phosphorus has attracted interest among both aquatic and soil ecologists. Photosynthetic organisms must assimilate mineral nutrients for growth which will be regenerated by herbivorous organisms. More recently it has been recognized that to a large extent bacteria depend on substrates which are poor in mineral nutrients, such as dead tissue or secretions from vascular plants. In such cases the primary decomposers assimilate rather than regenerate mineral nutrients, so that the availability of mineral nutrients may become limiting for decomposition as well as for primary production. It has thus been suggested that protozoa have a special role in the regeneration of mineral nutrients otherwise bound in bacterial biomass. Here it is sufficient to note that nitrogen and phosphorus cycling is closely coupled to carbon cycling in phagotrophic organisms. This is because these organisms (in contrast to bacteria or photosynthetic organisms) cannot assimilate inorganic nitrogen or phosphorus from the environment. Also the C:N and C:P ratios of phagotrophs are nearly invariant and in most cases are not very different from that of their food particles. The C:N ratio of protozoa is within the range of four to six (Finlay & Uhlig, 1981) which is comparable to that of bacteria. Large algal cells, however, may have somewhat lower levels of mineral nutrients with C:N ratios within the range of six to ten (Wheeler, 1983).

If the level of mineral nutrients in a phagotroph and its food particles are identical, the quantity of elements mineralized simply corresponds to the amount of mineralized carbon produced, that is, the amount of respiratory $CO_2$ divided by the C:N or the C:P ratio. In addition, an amount of organic nitrogen and phosphorus corresponding to the sum of the egested and excreted material is lost at each trophic level. Thus, in the idealized bacterium-flagellate-ciliate food chain discussed above, the two groups of phagotrophs will each mineralize about 30 percent and egest or excrete another 30 percent of the nitrogen and phosphorus they ingest, so that at the ciliate level, altogether about 85 percent of the bacterial nitrogen and phosphorus will have been released to the environment in either a mineral or an organic form. If the food particles contain less nitrogen or phosphorus, the amount of released mineral nutrients will be correspondingly lower.

How large a part of the total production of organic carbon is actually channelled through protozoa in the food chains of different ecosystems? In terrestrial ecosystems the most important primary producers are the vascular plants. It is a fact that a large fraction of this production is not utilized by herbivorous animals, but is decomposed in soil and litter by bacteria and fungi. The relative roles of these two components are not very well known and they undoubtedly vary according to environmental conditions. The fungi are predominantly consumed by various invertebrates, mainly nematodes, annelids, and arthropods. Bacterial populations, on the other hand, are mainly controlled by protozoan grazing.

In aquatic sediments, especially of shallow waters, a similar process occurs. The largest part of the organic production depends on the import of organic material which is primarily decomposed by bacteria which then serve as food for protozoa. In addition, the organic production of epibenthic microalgae and cyanobacteria is consumed to a large extent by protozoa. Until recently it was held that the carbon flow of marine plankton is predominantly based on a grazing food chain, in which large phytoplankters, like diatoms and dinoflagellates, serve as food for zooplankton organisms, which serve in turn as food for fish (see e.g., Steele, 1974). This view has been modified during the last decade. It is now believed that the largest fraction of the primary production is due to smaller eukaryotic algal cells and to cyanobacteria which are primarily consumed by protozoa. Also, a large fraction of the primary production is apparently excreted as dissolved organic materials, which provide the basis for a substantial bacterial production: Here again protozoa represent a necessary link in the food chain between bacterial cells and larger zooplanktonic organisms (Azam et al., 1983; Ducklow, 1983).

## 7.3    Patchiness and successional patterns

All natural environments, including the water masses of the oceans, show temporal and spatial variations or patchiness. Because of this heterogeneity, growth rates and life cycle characteristics are important components of the ecological niches of the organisms involved. Suppose a protozoan population and a population of a larger metazoan species live in the same environment and exploit a common food resource, the abundance of which fluctuates through time. The protozoan has a high potential growth rate and its population can trace the fluctuations of the food resource. During periods when the food resource is rare, protozoa may become rare or survive in an encysted state. The generation time of the larger metazoan species, on the other hand, may be long relative to the time periods of the fluctuations of the food resource and maintenance of this population will depend on the average density over time of the resource. In a similar way, spatial patchiness of a resource is utilized in a different manner by a small organism than by a large one which will only experience an average resource density. Thus environmental heterogeneity allows for the coexistence of different species if their sizes or population growth rates are sufficiently different, an idea which can be formally proven (Levins, 1979). Environmental heterogeneity is an important aspect of protozoan ecology and it must be considered in any discussion of the adaptive significance of many protozoan characteristics (Sections 4.3 and 5.1).

The generation of spatial and temporal heterogeneity is often extremely complex and involves both environmental patchiness and ecological interactions (Levin & Segel, 1985; Okubo, 1980). Environmental spatial heterogeneity occurs for a multitude of reasons, such as the particulate nature of sediments and soils, the steep chemical gradients of oxygen and hydrogen sulfide found around decaying material, and the particulate material which may be suspended in open-water masses in seas and lakes.

How small can such habitat niches be? An estimate can be made in the simple case where the enhanced local growth rate has to be balanced by the loss of cells due to their movement away from the niche. This model was originally suggested in order to estimate the minimum volume of a water mass which can sustain a bloom of photosynthetic algae (a "red tide"); it is discussed in detail by Okubo (1980). In this case it is supposed that the growth rate within the patch is $\mu$ and is zero everywhere else, and that the motility of the organism can be described by the diffusion coefficient D. The critical minimum diameter of such a patch which just sustains the population can then be shown to equal $c(D/\mu)^{1/2}$, where c is a constant which depends on the geometry of the habitat patch and which takes a value somewhere between unity and ten. To make some reasonable assumptions, consider a small flagellate with a doubling time of two hours ($\mu = 10^{-4}$/sec) and D = 0.1 mm$^2$/sec (see Section 2.3). In this case, the minimum diameter of a microhabitat, in the form of a physically defined water mass, would be at least several centimeters. In the real world, effective habitat patches of protozoa may be considerably smaller and defined protozoan communities are known to occur in niches as small as on the scale of millimeters. This is due to behavioral responses to chemical or other stimuli (thus D is not invariant as assumed in the model) and to attachment to solid surfaces. In the light of the model the behavioral responses discussed in Section 2.3 can be seen as adaptations for exploiting small habitat patches in a spatially heterogenous environment.

The succession of protozoa on or in degrading organic material, such as detrital particles or carrion, adds a time dimension to niche heterogeneity. Laboratory observations of such successions in "microcosms" not only describe what presumably goes on under similar conditions in nature, but, as shown in Section 8.3, they allow the estimation of rates and efficiencies of growth and the transfer of materials. Such successions are probably characteristic of all natural environments. In soils, temporal patchiness is also generated by rainfall, which stimulates bacterial growth, followed by growth of bacterivorous protozoa (see Figure 10.1). In planktonic environments, seasonal changes induce algal blooms which again lead to the growth of protozoan populations.

The study of prey-predator systems (thoroughly discussed in e.g., May, 1973) gives insight into the behavior of natural communities even though the complexities of the real world rarely allow the use of simple models for a quantitative description. This is because the simple models do yield certain qualitatively definitive predictions. One such prediction is that if the growth conditions of the prey species are enhanced, the principle result will be an increase in the average size of the predator population. Thus the average population density of bacteria in seawater is approximately the same in the surface layers of oligotrophic oceans and in eutrophic coastal seas (0.5 to 2 × 10$^6$ cells/ml), the minimum level which will support the growth of bacterivorous protozoa. This observation supports the idea that bacterial populations are controlled by protozoan grazing (see Section 8.2).

Models of predator-prey systems show an inherent tendency toward coupled population oscillations. Different versions of the classical Lotka-Volterra models of predation will either have a globally stable point of equilibrium or show neutrally stable oscillations. These models, however, assume a linear, functional (and numerical) response of the predator to prey density, which is unrealistic. Models in which the functional response of the predator is hyperbolic (see Section 3.1) show a stable equilibrium and/or a stable limit cycle. In such oscillating predator-prey systems, the period of the predator lies about one-fourth of a cycle behind that of the prey and the period of the cycle is proportional to $(\mu d)^{-1/2}$, where $\mu$ and d are the maximum growth rate constant of the prey and the death rate constant of the predator, respectively. The tendency to oscillate and the amplitude of oscillations is a function of the ratio between the environmental carrying capacity for the prey species (in the absence of predation) and the average prey population (the minimum number which can sustain a predator population). Thus increasing the carrying capacity of the prey destabilizes the system. This phenomenon is termed the "paradox of enrichment" (May, 1972; Rosenzweig, 1971) and predicts that increasing the resources for organisms at the base of food chains will lead to increasing population fluctuations.

Simple, artificial protozoan communities which consist of only one prey and one predator species are well suited for illustrating such models, as was first done by Gause (1934) with *Didinium* and *Paramecium*, and with *Paramecium* and yeast cells (see also Luckinbill, 1973, 1974). Even laboratory microcosms which include several species frequently show coupled prey-predator oscillations, such as between bacteria and bacterivorous protozoa, for example (Figure 8.3,B). The basic mechanisms which the simple predator-prey models describe undoubtedly contribute to population oscillations observed in natural habitats. However, due to the complexity of most natural habitats (which results from environmental fluctuations, and the fact that the predatory species themselves are victims to other predators or may feed on several different prey species), observed population fluctuations can rarely be analyzed in terms of simple models. There are, however, some exceptions: Figure 8.3,A shows the coupled oscillations of suspended bacteria and bacterivorous flagellates in estuarine water and in this case, the quantitative behavior of the system can be predicted from the behavior of the organisms as studied in the laboratory. The amplitudes of the population fluctuations of this system varies in different waters as a function of productivity, and in very oligotrophic areas populations of flagellates and bacteria remain nearly constant (Fenchel, 1986b).

## 7.4  Niche differentiation and coexistence

The "competitive exclusion principle" is held as one (among few) general principles of ecology. It expresses the idea that, given a universe with only one resource, only one species population persists. Inspired by the mathematical formulation of this idea by Lotka and by Volterra, Gause (1934) was the first to demonstrate interspecific competition for common resources in experiments with different species of *Parame-*

*cium*. These experiments are widely quoted in textbooks as examples of competitive exclusion, and they have played a historical role in the development of ecology. In one sense, however, they are of limited interest. Gause deliberately chose the experimental conditions so that they closely mimicked the assumptions of the Lotka-Volterra equations describing two-species competition, and so it is not strange that the outcome accords with the prediction of the model.

Most interesting however (as is often true in such attempts to "prove" simple models), are the cases where the result of the experiment deviates from the predictions, since this discloses mechanisms not described by the model. Thus in competition experiments between *Paramecium aurelia* and *P. bursaria* using yeast cells as food, Gause found stable coexistence. The reason for this is that the former ciliate is a superior competitor for suspended yeast cells, whereas the latter ciliate can better utilize sedimented food particles on the bottom of the test tubes used as experimental universes. The experiment shows that even in this small and seemingly homogeneous environment subtle habitat differences may allow coexistence. Since the experiments by Gause, protozoa have not played a particularly important role in the study of competition and coexistence as compared to larger organisms (in particular birds and vascular plants). This is because the study of coexistence and resource sharing in nature requires knowledge of the natural history of the individual species. Such data are more easily obtained for birds than for protozoa.

In modern ecological theory the competitive exclusion principle has been combined with a theory of the ecological niche to allow a quantitative prediction of the differential utilization of common resources which are necessary for stable coexistence (the "theory of limiting similarity"; MacArthur & Levins, 1967). The niche is considered to have three dimensions: (food) resources, habitat, and time, and the differential exploitation of any of these dimensions may explain regional coexistence of different species.

In nature, food particles occur over a continuous size spectrum. This allows for several coexisting species, each of which exploits a certain size range. This is because all mechanisms for catching food particles are most efficient only for a restricted range of particle sizes. Optimum food particle sizes correlate with morphological features (such as beak dimensions or overall size in birds). An example of such a "guild" of coexisting, congeric species which are specialized on different food particle sizes is shown in Figure 7.1: Species of *Remanella* occur around the anoxic-oxic boundary layer in marine, sandy sediments and clearly have a differential utilization of food particles. (These species eat a variety of food items, in addition to diatoms, such as flagellates and other ciliates. However, only diatom frustules can be measured accu-

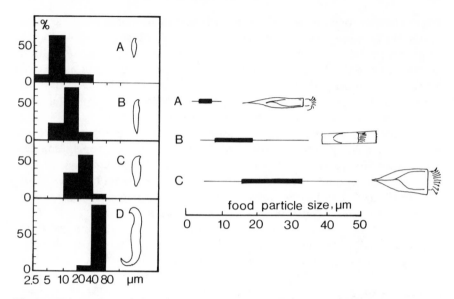

**Figure 7.1** Left: size distribution of ingested diatoms of four species of (frequently coexisting) species of the ciliate genus *Remanella* (drawn to scale). A: *R. margaritifera*, B: *R. rugosa*, C: *R. brunnea*, D: *R. gigas*. (Data from Fenchel, 1968.) Right: range of food particle sizes of three tintinnid ciliates: schematic drawing of each tintinnid is on far right. A: *Helicosomella subulata* (oral diameter: 20 μm), B: *Eutintinnus pectinis* (oral diameter 40 μm) and D: *Favella ehrenbergi* (oral diameter: 80 μm). (Unpublished data by Helene Munk Sørensen.)

rately in the feeding vacuoles of these ciliates.) This example is in accord with the prediction of MacArthur & Levins (1967) that food niches should be displaced by about one standard deviation to permit coexistence. A similar differentiation in food particle sizes of two coexisting species of *Loxodes*, a freshwater relative of *Remanella*, has also recently been demonstrated (Finlay & Berninger, 1984). It has also been found that in the surface layers of estuarine sediments the unrelated ciliates feeding on microalgae show limited overlap with respect to food particle sizes (Fenchel, 1968).

Figure 3.5 shows how filter-feeding ciliates differ with respect to the retention of particles of different sizes, and that even with regard to particles only of bacterial dimensions (0.2 to 2 μ), there is scope for coexistence of different species on the basis of particle sizes. This is mainly a consequence of mechanical constraints of food particle capture. Figure 7.1 shows the range of food particle sizes ingested by three species of coexisting tintinnid ciliates: The lower and upper sizes of these ranges are determined by the free distance between neighboring membranelles and the diameter of the lorica, respectively; and such constraints contribute to the diversity of these ciliates in the marine plankton. The role of the mechanical properties of food particles is also reflected in the structural diversity of the "pharyngeal basket" of different cyrtophorid ciliates (Figure 3.7,D; Section 3.2); this example is reminiscent of different types of beak morphology in birds.

Specialization to certain types of food on the basis of properties other than mechanical ones undoubtedly occurs, at least among raptorial ciliates. It has been established that *Didinium* specializes on species of *Paramecium* as food while some predatory ciliates seem to avoid hypotrich ciliates as food (Dragesco, 1962). The exact functional basis for such examples is not known, but the phenomenon will allow for the coexistence of several carnivorous ciliates within a microbial community.

There are innumerable examples of habitat specializations in protozoan species which explain regional coexistence; many will be discussed in Chapters 8–11. Habitat preferences occur on different scales; many species of radiolaria are confined to certain broad depth zones which extend over hundreds of meters. At the other extreme, a variety of sessile ciliates living on gammarid crustaceans have habitat niches measuring only a few millimeters, since some species only occur on the antennae, some only on the gills, and some only on the pleopods, etc. (see Figure 11.1). As discussion in Section 7.3, attached forms have a potential for exploiting smaller habitat niches than do free-swimming ones, which tend to leave very small patches due to random motility. The fact that zonation patterns of motile protozoa within a scale of a centimeter or less may be found in chemical gradients (see Figure 8.8) shows the significance of chemosensory behavior.

The differential exploitation of the time-correlated occurrence of resources is a complex subject. Traits which are related to time niches are discussed in Sections 4.3 and 5.1, and several ecological aspects were mentioned in Section 7.3. Examples of time niches also include the seasonal occurrence of protozoa which are encysted (or very rare) during other parts of the year; several such examples are mentioned in Noland & Gojdics (1967) for freshwater protozoa, but neither the mechanism nor the ecological significance is understood in detail. The concept of "competition-extinction equilibria" implies regions with resources which are unevenly distributed in time and space. When such a resource patch appears it may be colonized by various protozoa. Within a single patch some species will prove to be inferior competitors and eventually become locally extinct due to resource competition. These inferior competitors, however, may still persist regionally if their rates of reproduction and dispersal are sufficiently high to ensure the colonization of new resource patches before they become extinct in the old one. Theoretical models of such competition-extinction equilibria are found in Levin & Paine (1974).

Protozoa and protozoan communities offer a unique possibility for the study of time-related aspects of ecological niches due to the short generation time involved. Some authors (e.g., Luckinbill, 1979; Lütenegger et al., 1985; W.D. Taylor, 1978) have attempted to classify various ciliated protozoa as either superior competitors or "fugitive" species on the basis of the concept of "K and r selection" (MacArthur & Wilson, 1967). The basis of this idea is that some organisms specialize by sacrificing competitive ability for a high reproductive potential and a high capacity for dispersal. These attempts are not entirely successful since they mainly consider the maximum growth rate as a correlate of fugitive existence; this may be incorrect from a theoretical point of view. In addition, the measured maximum growth rates will be a function of the specific experimental conditions used and they will, under all circumstances, be correlated with cell size. The studies also tend to ignore the possibility that the food niches of the compared species may be rather different. A deeper understanding of time-related components of the niche requires that aspects of polymorphic life cycles (including the ability to form cysts, special dispersal stages, etc.) are related to the ecology of the species in nature and so far this has only been done in a very general way.

It is a common observation that within any single taxon of animals there are many small and few large species. This is rationalized by the fact that the degree of environmental, spatial, and temporal patchiness in an area depends on the size of the animals. What appears to be an homogeneous forest to an elephant is a universe of landscapes to a small rodent. There are therefore many more habitat and time niches available to small than to large organisms. May (1978) illustrated this by comparing the numbers of terrestrial animal species of different size classes over the range from 0.5 millimeters up to the size of an elephant. Actually most species (more than half a million) measure from 0.5 to 1 centimeter while numbers of even smaller species tend to decline. May (op. cit.) suggests that the dearth of very small species is because of inadequate species taxonomy for the very smallest organisms. The view that the increasing number of habitat niches primarily accounts for the differences in numbers of small and large animals is probably correct. However, the declining numbers of species among the classes of the smallest animals may not only be due to inadequate taxonomy; it may reflect a real phenomenon. This interpretation receives support if protozoan species numbers are taken into account. Comparisons with numbers of terrestrial animals is perhaps unfortunate, since the large number of insect species (about three-quarters of a million have been described) is probably explained by the diversity and spatial complexity of vascular plants. Nevertheless, the total number of protozoan species (even if the incomplete species taxonomy of many groups is taken into account and even if we allow for the fact that many species remain undescribed) is in fact not very impressive, when compared to the total number of invertebrate groups of animals.

According to Corliss (1984) at most some 30,000 species of recent, phagotrophic protists have been described. Among these species, about 8,000 are ciliates; 1,100 are rhizopods (of which most are foraminifera); 5,000 are actinopods (mostly radiolaria, but Anderson (1983) believes that this number is inflated); and perhaps 3,000 to 4,000 are flagellates; these numbers include some parasites and also some predominantly photosynthetic forms. In contrast, even for such relatively homogeneous groups as nematodes and copepods, 10,000 and 4,000 species, respectively, have been described. It is unlikely that the inverse relationship between species numbers and body length found by May (1978) for animals longer than 0.5 centimeters can be extrapolated to organisms in the size range of protozoa (this relationship would predict that there should be perhaps $10^7$ to $10^8$ species of protozoa).

There may be several different reasons for the relatively low number of protozoan species. One is related to the topic of the following section: small organisms tend to have a wide geographical distribution and allopatric speciation is accordingly less important. Also, climate may play a smaller role for small organisms; in microhabitats such as tiny ponds in a temperate climate, the yearly span in temperature may compare to that of average temperatures from tropical and arctic regions combined. Also the food resources of protozoa (bacteria, microalgae, other protozoa) are, to a large extent, cosmopolitan. Finally, the behavioral complexity of larger invertebrates as compared to protozoa may allow the former to exploit more specialized or rarer habitats.

## 7.5 Biogeography of protozoa

The way in which the distribution of organisms in time and space can be explained depends on the scale involved. At scales comparable to a generation time and to the distance within which an individual may move around, distribution patterns can be the result of ecological factors, such as the local physical or chemical environment, food resources, interspecific competition, or demographic or environmental stochasticity. At the other extreme, namely, an evolutionary time scale and large distance, which constitute efficient migration barriers, explanations of distribution patterns may invoke historical (e.g., continental drift) or evolutionary events. At intermediate scales, a variety of factors, such as climatic and seasonal weather patterns, ocean currents, or mechanisms which facilitate long-distance dispersal, may be responsible for the observed distribution patterns.

It is a conventional view that most protozoan species have a cosmopolitan distribution. Symbiotic protozoa associated with metazoan species are often very host specific and constitute an obvious exception. With regard to free-living forms it seems that in similar habitats, identical species occur everywhere on earth. However, there are several reasons for modifying this generalization. First of all, faunistic surveys of protozoa are much more extensive in Europe, in North America, and in the adjacent seas, than anywhere else. The varying degree of taxonomic resolution at the species level (see Section 1.4) in conjunction with the authority of textbooks and taxonomic memoirs may bias the picture. The protozoologist who collects protozoa in a new area tends to identify species as being the same as forms already described and named from Europe, for example, even though subtle differences may occur. Also the apparent absence of a species may simply signify that it has not yet been found in a particular area.

In fact there are examples of biogeographical patterns in the distribution of protozoan species. The giant amoeba, *Chaos carolinensis*, known from freshwater habitats in North America, has apparently never been found in Europe and it is very unlikely that it has been overlooked. Dragesco (summarized in Dragesco & Dragesco-Kerneis, 1986) has studied freshwater and marine ciliates in Africa where several novel species were found which were endemic to that continent or to tropical regions, and which could not possibly have been overlooked in more intensely studied areas. Among soil protozoa, at least the testate amoebae show definite biogeographical provinces (e.g., South America; see Bonnet, 1979). The tropical and subtropical distribution of giant shallow-water foraminifera have already been discussed (see Section 6.2).

The geographical distribution of the sibling species belonging to the *Tetrahymena pyriformis* complex (Elliott, 1973) is particularly interesting. Some of these species are endemic to Europe or to North America although there is some overlap in the species compositions, but South America (including some Pacific Islands) and Australia each have their own species. It is tempting to infer isolation and subsequent speciation due to continental drift as an explanation for this pattern. The fact that these sibling species, despite their morphological similarities, are very different at the molecular level (see Section 1.4) suggests that also in other cases, very similar protozoa from different areas may not be very closely related, but that their similar phenotypes represent an adaptive evolutionary peak.

The most peculiar example of an endemism of freshwater protozoa is that reported for Lake Baikal by Svarchevsky and by Gaevskaja (reviewed in Dogiel, 1965). Some of these forms are ectocommensals on

endemic gammarid crustaceans. However, a number of strange planktonic ciliates were also described and some of them represent endemism
at the family or generic level: Further studies on this fauna would be
of interest. The geographically most confined endemism recorded seemed
to be a ciliate belonging to the genus *Condylostoma* isolated from a
140-meter-long saline lake on the coast of the Sinai peninsula (Wilbert
& Kahan, 1981). More recently, however, this about two millimeter
long ciliate have been recorded from the west coast of Africa (Dragesco
& Dragesco-Kerneis, 1986).

In spite of these examples it is still true that protozoan species from
freshwaters, soils, or coastal seas in general, are widely distributed as
compared to larger metazoan species which at least are usually endemic
to continents, parts of continents, or even to mountain ranges or small
islands. One reason for this difference is the fact that most protozoa
have a high potential for dispersal. Cysts which resist desiccation can
be spread by the wind over great distances, but even species which do
not form such cysts can be transported over long distances on or in
aquatic birds and insects, or on the feet of mammals (see e.g., Maguire,
1963; Revill et al., 1967; Schlichting & Sides, 1969). Marine species
can migrate along coasts or may be transported over long distances and
even cross oceans with disrupted seaweed or in small amounts of sediment trapped in floating logs or other material. Direct evidence for the
dispersal ability of protozoa also exists. Maguire (1977) studied the
colonization by freshwater protozoa of containers of water placed on
the volcanic island Surtsey, which emerged from the sea some 30 km
south of Iceland in the 1960's. A remarkably rapid appearance of a
number of species was found. Inland saline waters which originated
from salt mining in Northern Germany contain a normal fauna of marine
ciliates (Kahl, 1928).

Cairns & Ruthven (1972) studied the fauna and migration rates of
freshwater protozoa on a small island in the Bahamas having a very
restricted amount of natural freshwater. They found that the rate of
colonization of open containers with sterile water was slow as compared
to similar experiments on the mainland. They discussed their results in
terms of the "theory of island biogeography," which is discussed below
(MacArthur & Wilson, 1967), and accordingly expected that the species
number on the island would equilibrate at a lower level than in similar
habitats on the mainland. This may have been the case, but the findings
are not so easy to interpret because many soil protozoa also occur in
freshwater and so the pool of local species which potentially could
colonize the experimental containers might have been larger than
expected.

The theory of island biogeography does, however, throw some light
on the wide distribution of protozoan species. According to this theory,

the species number of any isolate (an island, lake, a mountain top) is to be understood as a dynamic equilibrium between the immigration of new species and the local extinction of previously established species. The process of extinction may be accelerated by competition, predation, or environmental stochasticity, but eventually extinction is probably due to demographic stochasticity in small populations. Therefore the equilibrium number of species within any taxon is lower in small isolates (that is, "islands" in a broad sense including e.g., ponds and lakes) which have smaller absolute population sizes. However, since protozoa are small, their absolute population sizes may be immense, even in a small isolate like a pond and even if the species in question would seem relatively rare. The probability of a local extinction of a species population of protozoa is therefore very small as compared to the extinction probability of a population of birds on a small island. The theory of MacArthur & Wilson therefore predicts that the "distance effect" (that is, the effect on species richness due to decreased migration rates over larger distances between isolates) will be slight. One fundamental reason for the wide and sometimes global distribution of protozoan species (and other small organisms) is therefore the huge absolute population sizes attained by these organisms. This explanation is appealing because it does not imply any special properties of small organisms (such as a slow rate of evolution or an extraordinary potential for dispersal) as suggested by some authors, but is only based on the fact that small organisms form populations with a large number of individuals. This interpretation predicts that larger protozoa would show a higher tendency to form biogeographical patterns than small ones. The giant amoeba endemic to America and the predominantly large, endemic African ciliates discussed above may be consistent with this, but it is not yet possible to make generalizations.

The geographical distribution of oceanic, planktonic forms remains to be discussed. In the case of large, skeleton- or lorica-forming groups (tintinnids and in particular foraminifera and radiolaria) this distribution has been studied in some detail; this subject was recently reviewed by Zeitzschel (1982) for the previously mentioned groups and for the radiolaria by Anderson (1983). Biogeographic assemblages of species are characteristic of physically defined water masses such as the major oceanic currents and circulation gyres. The most important zones are defined by five circumglobal belts: a northern and a southern cold-water belt, a northern and a southern subtropical belt, and a tropical belt. These zones basically describe the faunal assemblages formed by the planktonic foraminifera, of which there are less than forty species, most of them occurring in the warm belts. Their distribution is probably mainly governed by temperature and the occurrences of tests of the different species in sediment cores are therefore often used as paleoclimatic indicators.

The biogeography of radiolaria seems to be more complex; some species have a global distribution while others are confined to climatic zones or to certain oceanic circulation gyres or upwelling zones. Some cold-water species show a bipolar distribution. This is explained as a result of the pleistocene glaciations during which the distribution of the species populations could have been continuous.

# 8

# Protozoan Communities: Marine Habitats

## 8.1  Introduction

To most ecologists the term "biotic community" means an assemblage
of species populations delimited in time and space within which the
populations interact. In practice it is not possible to use a very strict
or precise definition of communities when discussing taxonomically
defined assemblages in nature. It is obviously necessary to include prey,
predators, and competing species which are not protozoa. On the other
hand, it is likely that many protozoan populations which occur together
in time and space interact only weakly or not at all. Thus the largest
and the smallest protozoan species of many habitats are not likely to
engage directly in prey-predator relationships or to compete for common resources and any possible ecological interactions will be weak
and indirect. Also the spatial and temporal boundaries of biotic communities are rarely, if ever, sharply defined.

Another problem is one of sampling scale. A bottle of seawater may
be considered a sample of a pelagic community, but it is likely to contain
suspended particles with distinct assemblages of adhering protozoan
species otherwise typical of detrital sediments. Likewise, a sediment
sample may include several distinct protozoan communities. Failure to
recognize this is one reason why the structure of many protozoan communities is incompletely understood.

We combine here the approach initiated by Picken (1937) and Fauré-
Fremiet (1951), namely, to look at the mutual interactions of individual
species populations, with more quantitative considerations of the role
of protozoa in different habitats. Each subsection of this and the fol-

102

lowing two chapters will discuss a different community or habitat; the considerations above explain why the classification of habitats may, in some cases, seem quite arbitrary.

## 8.2   Marine pelagic protozoa

**General considerations**    The diversity and abundance of marine planktonic protozoa has long been recognized, as witnessed in Haeckel's (1887) memoir on the radiolaria collected during the Challenger expedition. Another early study is that of Lohmann (1911) who discussed the significance of small, heterotrophic protists in seawater. However, it is only during the last decade that the importance of protozoa in pelagic food chains has been fully recognized and this aspect has recently enjoyed an increasing interest among marine ecologists.

In part this reflects a changed picture of pelagic food chains in general. It is established that a large fraction of the primary production in the sea is due to minute eukaryotic algae and to unicellular cyanobacteria which are too small to serve as food for most metazoan zooplankters such as copepods. It has also been found that a substantial fraction of the photosynthetically produced organic matter is lost to the environment as dissolved material. The reason for this is not clear, but as much as 50 percent of the primary production seem to be channelled via dissolved organics to bacteria, which again are responsible for a production of particulate organic material corresponding to 20 to 30 percent of that of photosynthetic organisms. This bacterial production seems almost entirely to be consumed by small protozoa, which in turn serve as food for larger protozoa and metazoan zooplankters. This aspect of pelagic food chains has been termed the "microbial loop" (Azam et al., 1983; see also Ducklow, 1983, and Williams, 1981). The succession of bacteria, heterotrophic flagellates, and ciliates following phytoplankton blooms was first observed and interpreted in upwelling areas in the open sea by Sorokin (1977, 1978). A schematic representation of a pelagic marine food chain is shown in Figure 8.1. The essence of this new paradigm of planktonic marine food chains is that there are a higher number of trophic levels (mainly represented by microscopic phagotrophs) than hitherto believed. As a consequence a considerable fraction of the reduced carbon must be mineralized before it reaches large organisms such as copepods and fish. The problem of whether current estimates of the primary production of the sea can account for this has hardly been addressed yet, but the existence of the microbial loop suggests that the photosynthetic productivity of the plankton has hitherto been underestimated.

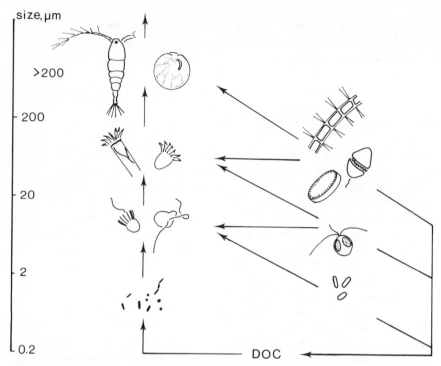

**Figure 8.1** The "microbial loop" of a marine planktonic food chain. (DOC: dissolved organic carbon; for further explanation see text.)

In Figure 8.1 the size groups of organisms are classified according to Sieburth (1979) into "picoplankton" (0.2–2 μm), "nanoplankton" (2–20 μm), "microplankton" (20–200 μm), and "mesoplankton" (>200 μm). This classification is useful since it reflects the fact that a length ratio of ten roughly represents one step in a food chain (see Section 7.2). Also these size classes tend to include some taxonomic information as well. Thus the picoplankton mainly includes heterotrophic bacteria and cyanobacteria; the nanoplankton, mainly heterotrophic flagellates and various groups of photosynthetic flagellates; the microplankton includes the ciliates, most diatoms, and photosynthetic dinoflagellates; and finally the mesoplankton includes the largest heterotrophic protozoa (mainly radiolaria and foraminifera), in addition to metazoan zooplankters. There are, of course, several exceptions to these generalizations; some ciliates are smaller than 20 μm and different species of heterotrophic dinoflagellates are represented among the nano- and micro-, as well as the mesoplankton. This size classification is, however, valuable as a framework for a more detailed discussion of protozoan plankton.

Planktonic communities are spatially structured due to hydrodynamic phenomena such as gyres and ocean currents and other large-scale physical isolations of water masses. The question of whether microbial planktonic communities also show a spatial structure on a much smaller scale has recently been discussed. It has been demonstrated that particulate aggregates ("marine snow") and sinking detrital particles harbor distinct microbial communities consisting of bacteria and bacterivorous protozoa, such as flagellates, amoebae, and ciliates, some of which are not typical planktonic forms (Caron et al., 1982; Silver et al., 1984). Direct counts of bacteria and protozoa (based on fixed samples filtered through membrane filters and enumerated under the fluorescence microscope) suggest that in fact only a small fraction of the bacteria in oceanic water is associated with particulate material. Many heterotrophic flagellates, however, do have mechanisms for temporary or permanent attachment and they are often associated with suspended particulate material in at least the more eutrophic situations (Fenchel, 1986b). The rapid turnover of microbial populations and of mineral nutrients in oligotrophic waters has inspired some authors to consider whether microbial populations predominantly occur in some sort of microaggregates, which could explain this phenomenon under conditions of very low concentrations of nutrients in bulk water samples. Goldman (1984) considers the role of microbial communities consisting of primary producers, bacteria, and microprotozoa associated with particulate matter and shows that such aggregates allow for a more rapid population turnover. He also suggests that these aggregates, to a large extent, remain undetected because they fall apart when exposed to the usual methods for enumerating microorganisms in water samples. Mitchell et al. (1985) consider microzones surrounding individual phytoplankton cells in which increased concentrations of dissolved organic material attract bacteria through chemosensory mechanisms. These authors conclude, however, that for realistic values of diffusion rates, sinking rates of particles, and turbulence, such microaggregates can only be maintained under special conditions. The role of a spatial structure of microbial populations at a microscopic scale is therefore still not clear and the suggested models are very hard to test, given the current methods for studying microbial plankton.

**The heterotrophic nanoplankton**    The pelagic protozoa within the size range of 2 to 20 $\mu$m includes amoebae, flagellates, and some ciliates. The latter group will be discussed in the following section. Pelagic amoebae have received little attention. Amoeboid cells are frequently associated with suspended, detrital particles and it is also likely that in shallow waters, amoebae in the sediment occasionally become suspended. Davis et al. (1978) isolated and enumerated amoebae (using a serial dilution method) in water samples from the North Atlantic. Only a few cells per liter of water were found. Samples of the surface microlayer, however, yielded nearly a thousand-fold-higher concentration of cells, so it is likely that amoebae play a role in the ecology of the surface film of natural waters; this film is also able to concentrate bacterial cells.

Flagellates constitute the dominant component of the heterotrophic nanoplankton as regards diversity, numbers, and biomass. Some of these forms were described in considerable detail by Griessmann (1914) with the light microscope. However, for many years taxonomic emphasis was restricted to forms which carry silicious skeletons or spines and which can be studied as whole mounts in the transmission electron microscope, namely, acanthoecid choanoflagellates and some chrysomonads (see Figure 8.2). The diversity and taxonomy of the numerous forms which do not possess such mineral structures is still incompletely known; references to the literature on the taxonomy of marine microflagellates are given in Fenchel (1986b) and Fenchel & Patterson (1986).

The single most important group of marine plankton is undoubtedly the choanoflagellates; they often account for as much as fifty percent of the heterotrophic flagellate fauna in water samples. Some forms are permanently attached to suspended particles or algal cells, but most are free-swimming. Among the three families of choanoflagellates, the Codonosigidae (Figure 8.2,A,C) are naked, the Salpingoecidae are encased in a membranous collar, while the Acanthoecidae possess a complex, silicious skeleton (Figure 8.2,E). Colonial forms occur within all three groups. The 3- to 5-μm-large cells possess a single, smooth flagellum capable of driving water through a tentacular collar which acts as a filter (Figure 3.4,A). In particular, the acanthoecids include a surprisingly large number of species.

While the coanoflagellates are never pigmented, the chrysomonads represent a continuum extending from phototrophs to obligatory phagotrophs and some simultaneously possess a chloroplast and ingest particulate food. The genus *Paraphysomonas* comprises nonpigmented forms with silicious spines (Figure 8.2,D) while naked, nonpigmented forms are assigned to the genus *Monas* (Figure 8.2,B). The identity of the latter is probably often confused with *Ochromonas* because the chloroplasts of the latter may be very hard to observe under the light microscope. All three genera are common in seawater and all can drive water currents toward the cell with their hairy flagellum; the cell intercepts and ingests particles which touch the cytostome (Figure 3.6,A). The bicoecid flagellates are related to the chrysomonads: They are never pigmented and the smooth, second flagellum is longer than in the chrysomonads and attaches to solid surfaces, water films, or to the bottom

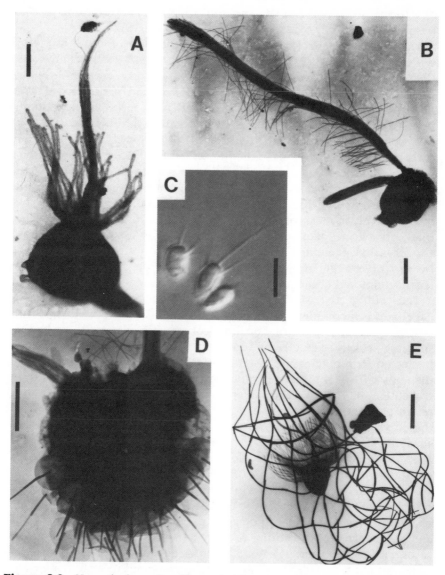

**Figure 8.2**   Nanoplankton flagellates. A. The choanaoflagellate *Monosiga*. B. The chrysomonas *Monas*, showing the hispid and the smooth flagellum. C. The colonial choanoflagellate *Desmarella*. D. The chrysomonad *Paraphysomonas vestita*, with silicious spines. E. The choanoflagellate *Diaphanoeca grandis*, showing the complex, silicious lorica and the tentacular collar. A, B, D, and E are whole mounts of fixed cells photographed in the transmission-electron microscope; C shows living cells. Scale bars: A, B, D: 1 μm, C, E: 5 μm.

of a lorica built by the cells. The nonloricate *Pseudobodo* may be extremely common in seawater samples (Fenchel, 1982a).

The helioflagellates show affinities with the chrysomonads as well as with the heliozoa. *Pteridomonas* and *Actinomonas* use a stalk to attach temporarily to solid surfaces. A hairy flagellum drives water through a pseudopodial collar which strains water for food particles. In *Ciliophrys*, feeding resembles that in a typical heliozoan: the flagellum is immobile and motile bacteria or other small organisms are caught on the pseudopodia which radiate out in all directions. The pseudopodia carry extrusomes which make the prey cells stick. Only when disturbed, will a cell activate its flagellum and swim away (Davidson, 1982; Patterson & Fenchel, 1985). Various other groups of flagellates are also represented among the heterotrophic nanoplankton, but they play a quantitatively modest role. Colorless euglenids and bodonids occur regularly in marine plankton, but they are usually associated with solid surfaces; nonpigmented cryptomonads are also found, as are some *Tetramitus*-like flagellates, and some other forms whose taxonomic affinity is still unknown (Fenchel & Patterson, 1986).

The ecological basis for the relatively high diversity of planktonic flagellates is incompletely understood. Particle-size selectivity undoubtedly plays a role. Thus the choanoflagellates are adapted to utilize the smallest prokaryotic cells, as shown by the free space between neighboring collar tentacles and also by the fact that the lorica of some acanthoecids will only allow very small particles ($<0.5$ $\mu$m) to pass on to the collar. In contrast, the collar of *Pteridomonas*, for example, will retain only particles larger than 1 to 2 $\mu$m. The raptorial feeders like the chrysomonads can, in principle, intercept particles of a wide size span, but the efficiency is likely to vary inversely with food-particle size. *Paraphysomonas* will ingest and grow on bacteria as well as on algal cells approaching its own cell diameter (Caron et al., 1985). It would seem that other mechanisms which lead to niche diversification play a role, but this is still speculative. Thus it is likely that differing life-cycle characteristics occur in response to environmental variation (for example, in terms of the ability to encyst in response to declining food resources; see Sections 4.3 and 5.1), and the amount of suspended particles (as solid surfaces to which some forms tend to attach) may also play a role in niche diversification. Whether habitat diversificiation occurs as a function of different water depths is not known. In addition, some species seem to occur in large numbers only during certain seasons.

Several microflagellate species have been isolated in pure culture (by using serial dilution of seawater samples and bacterial suspensions as food) and in general it is the quantitatively most abundant forms which are obtained (Fenchel, 1986b). The establishment of such cultures has allowed bioenergetic parameters of the organisms to be studied in the laboratory (Caron et al., 1985; Goldman et al., 1985; Fenchel, 1982b; Sherr et al., 1983; see also Figures 4.1 and 4.3 and Chapter 4). The flagellates are typically capable of clearing food particles from a volume of water which is about $10^5$ times their own cell volume per hour; that is, between $2 \times 10^{-6}$ to $2 \times 10^{-5}$ ml/h depending on size.

Microflagellates can be quantified in seawater samples by various methods. The most successful one is to filter fixed samples through membrane filters, stain the filtered cells with a fluorochrome dye, and count them under the fluorescence microscope. This makes it possible to distinguish photosynthetic forms from nonpigmented ones (due to the red autofluorescence of chlorophyll), to identify them (at least to the generic or family level), and to simultaneously count bacterial cells (Davis & Sieburth, 1982; Fenchel, 1982d). In this way, a substantial amount of data on the quantitative occurrence of phagotrophic flagellates in seawater has been obtained (Fenchel, 1982d, 1986b; Sherr et al., 1984; Sieburth & Davis, 1982, and references therein). Typically around a thousand flagellates per milliliter of seawater are found, that is about one flagellate per thousand bacteria. Thus, bacteria and their predators represent similar biomasses, as predicted in Section 7.2. Numbers of flagellates are lower in oligotrophic ocean water and higher in eutrophic coastal waters: In offshore waters their numbers decrease with depth. While bacterial and flagellate numbers remain relatively constant in oligotrophic waters, the populations show fluctuations in eutrophic ones. Such fluctuations are triggered by phytoplankton blooms which lead in turn to increased bacterial growth. These fluctuations show complex patterns due to consumption of the flagellates themselves and to movements of water masses. In some cases, however, coupled oscillations between bacterial and flagellate populations may persist for several weeks (Figure 8.3,A). The period and amplitude of the oscillations are consistent with simple models of prey-predator systems and the growth parameters derived from pure culture experiments.

Data from clearance of bacteria by flagellates in pure culture applied to their numbers in nature show that in coastal seas during the summer, flagellates filter from 10 to >100 percent of the water column for bacterial cells per day. Such figures are consistent with contemporary estimates of growth rates of suspended bacteria in the sea and suggest that heterotrophic flagellates are the dominant agents controlling bacterial populations. This picture is supported by "microcosm experiments" such as the one illustrated in Figure 8.3,B. This figure shows the development of a freshly collected seawater sample in which larger planktonic organisms (>10 $\mu$m) have been removed, and a parallel sample in which flagellates have also been removed. Such experiments permit independent estimates of average clearance by flagellate populations. They also show that the minimum bacterial concentration which sup-

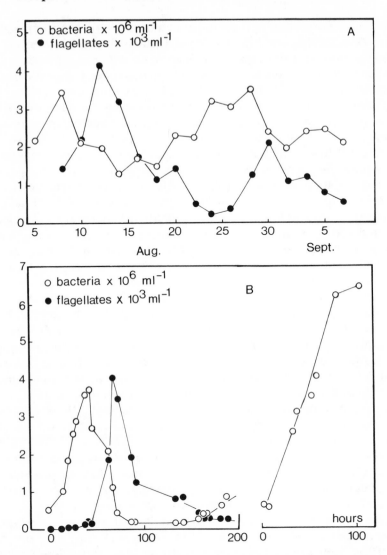

**Figure 8.3**   A. Fluctuations in numbers of suspended bacteria and heterotrophic nanoplankton flagellates in the water column of Limfjorden, Denmark, over a one month period in 1981. B. Changes in numbers of bacteria and flagellates in a freshly collected seawater sample (left) and of bacteria in a parallel sample from which flagellates were previously removed (right). (A: data from Fenchel, 1982d; B: data from Andersen & Fenchel, 1985.)

ports flagellate growth is around $10^6$ cells/ml. Related experiments in which organic material (such as decomposing phytoplankton cells or mucilage from kelp) is added to seawater samples illustrate the microbial food chain as a succession of events: Rapid growth of bacteria is followed by an increase in the flagellate population; thereafter, the bacteria decline in numbers; finally, flagellate numbers decline as bacterial numbers decrease and as ciliate populations increase in size (Linley et al., 1981; Newell & Linley, 1984).

**Microplankton protozoa**    The pelagic protozoa between 20 and 200 μm in length are predominantly ciliates, among which the almost radially symmetrical oligotrichs are the most numerous. Among these oligotrichs, the tintinnids have attracted the most attention because their loricae remain recognizable after collection by plankton nets and preservation in fixatives traditionally used for more robust plankton forms. Also, the lorica seems to facilitate taxonomic work and a large number of species have been described on the basis of its structure. However, the study of Laval-Peuto (1981) demonstrates that at least some species are polymorphic with respect to lorica structure and the high number of species described may not be realistic. Naked oligotrichs, in particular those represented by the genera *Strombidium, Tontonia, Laboea*, and *Lohmanniella*, are sometimes more numerous than are tintinnids, but their distribution and species taxonomy is still incompletely understood. Some typical marine oligotrichs are shown in Figure 8.4.

The oligotrich ciliates ingest food particles which are relatively large, but since these species as a whole represent a wide size range, so do their food particles. However, bacteria are probably not an important food for oligotrich ciliates. The small *Lohmanniella spiralis* retains particles down to 4 to 5 μm while the large tintinnids ingest particles up to 40 μm or so; thus their food consists of nanoplankters or in the case of the larger species, of dinoflagellates, diatoms, and other ciliates (Heinbokel, 1978; Rassoulzadegan, 1982; Rassoulzadegan & Etienne, 1981; Spittler, 1973).

Generalizing from the literature, there is typically just over one ciliate per ml of seawater. However, in eutrophic coastal waters and in upwelling areas the numbers show considerable fluctuation and several hundred oligotrichs (loricate or nonloricate) per milliliter are not uncommon; in some cases coupled fluctuations between nanoplankton organisms (photosynthetic and nonphotosynthetic forms) and ciliate populations are evident (Burkill, 1982; Hargraves, 1981; Ibanez & Rassoulzadegan, 1977; Rassoulzadegan, 1977; Sorokin, 1977, 1981). Few attempts to estimate the total grazing impact of oligotrich ciliates in marine plankton have been made. Heinbokel & Beers (1979) found that tintinnid grazing usually accounts for less than four percent, but some-

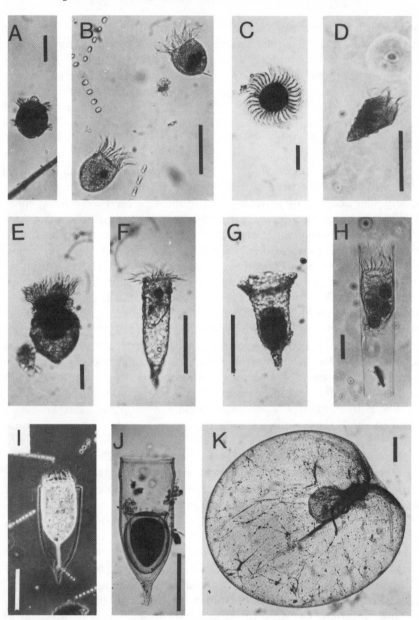

**Figure 8.4**  A-J: marine planktonic ciliates. A: *Didinium* sp. B-D: nontintinnid oligotrichs. B: *Lohmanniella oviformis*, C: *L. spiralis*, D: *Laboea strobila*. E-J: tintinnids. E: *Stenosomella ventricosa*, F: *Tintinnopsis cylindrica*, G: *T. campanula*, H: *Eutintinnus subulata*. I,J: *Favella ehrenbergi* (J: cyst). K: the giant dinoflagellate *Noctiluca miliaris*. Scale bars: A,D,F,G,I,J,K: 100 µm; B,C,E,H: 20 µm. (All specimens were collected in the Limfjord, Denmark, and photographed by Helene Munk Sørensen.)

times more than 20 percent of the daily, primary production off the California coast. This study did not consider the grazing effect on non-loricate oligotrichs, nor the grazing on heterotrophic organisms. Capriulo & Carpenter (1980) estimated that tintinnid grazing accounts for the daily removal of about 40 percent of the standing stock of chlorophyll in Long Island Sound, while Burkill (1982) estimated that around 60 percent of the primary production is consumed by ciliates (in particular tintinnids) in an estuary in southern England. Verity (1985) found that tintinnids ingested 16 to 26 percent of the annual net primary production in Narragansett Bay off the coast of Rhode Island and that the nitrogen excretion by these organisms supported 11 to 18 percent of the primary production.

Fenchel (1980d) argued that bacterivorous ciliates at most play only a modest role in marine plankton turnover, since their volume-specific clearance rate is lower (by a factor of about ten) than that of ciliates filtering larger particles and that of bacterivorous flagellates. Such ciliates are therefore likely to be competitive only where bacterial numbers are high, such as in eutrophic waters and in patches of decaying material. In general this view is confirmed by the literature; representatives of bacterivorous ciliates are rare or absent in quantitative surveys of marine plankton, although some small bacterivorous ciliates, such as *Uronema*, are often isolated from seawater samples. This observation may be due to ciliate cells being brought into suspension from sediments or associated with suspended particulate material. However, Sherr et al. (1986) found that in estuarine as well as in off-shore waters very small (around 10 μm) bacterivorous ciliates (mainly scuticociliates like *Cyclidium*) occur in large numbers, so they may have a significant impact on bacterial populations.

Finally, some ciliates have an obligatory association with photosynthetic organisms and apparently do not feed on particulate material. *Mesodinium rubrum* was discussed in detail in Section 6.2; it is not known for certain whether this species name actually includes more than one species or whether several related forms with a similar biology exist.

The quantitative role of heterotrophic dinoflagellates has attracted little attention. In oceanic waters their numbers seem to exceed those of ciliates; their diet includes bacteria, small flagellates, ciliates, and other dinoflagellates. Based on experiments on samples collected at various stations in the North Atlantic, Lessard & Swift (1985) estimated that the total population clearance by dinoflagellates is comparable to that of ciliates. Heterotrophic dinoflagellates and pigmented dinoflagellates, which are phagotrophs at the same time (e.g., *Oxyrrhis, Gyrodinium*), are also frequently observed in coastal waters; however, few studies exist on their ecological significance in the plankton. Smetacek

(1981) quantified heterotrophic dinoflagellates and ciliates in the plankton of Kiel Bight. The two groups represented comparable biomasses which had distinct maxima during early spring and autumn. Indirect evidence suggested that during the summer the protozoan populations were reduced due to predation by metazoan plankters. The best known and studied heterotrophic marine dinoflagellate is undoubtedly *Noctiluca miliaris* (Figure 8.4,K); although it reaches a size of 1 to 2 mm in diameter it is really a mesoplankton form. This species will ingest a variety of other planktonic organisms ranging from larger protozoa to crustaceans and fish eggs. For reasons not understood it shows seasonal mass occurrences in northern European waters (Sahling & Uhlig, 1982). During such episodes the cells tend to increase their buoyancy and where cells accumulate on the surface, their bioluminescence becomes obvious when the water is disturbed.

**The protozoan mesoplankton**    In addition to *Noctiluca* and its relatives, the protozoan giants of the plankton include the foraminifera, the acantharia, and the radiolaria (although some representatives of the latter are actually quite small). Aspects of the functional biology of these groups were discussed in Sections 3.2 and 6.2. In spite of their size, their esthetic appeal, and the fact that their structure and species systematics have been worked out in detail, surprisingly little is known of their biology and their ecological significance. There are several reasons for this: They are basically oceanic creatures and controlled cultures have not been obtained for any species. These delicate organisms lose their viability when captured with plankton nets, but even specimens collected individually in glass jars by divers can at best be kept only for a few weeks or months in the laboratory, provided with light and food. A completion of the life cycle and population growth has not yet been observed in the laboratory. Except for the mainly bathypelagic phaeodarians, they all possess photosynthetic symbionts in addition to their carnivorous habits. What is known about radiolarian ecology is reviewed in Anderson (1983). In surface waters, the solitary forms may occur at densities of several hundreds per cubic meter. In the surface layers of oligotrophic ocean water, colonial radiolaria may be the dominant macroplankton. Each colony may contain thousands of cells and measure up to five centimeters in diameter. Typically from 0.04 to 14 colonies per cubic meter are found, but sometimes they accumulate at much higher densities (Swanberg, 1983).

While the oceans harbor many hundreds of species of radiolaria, less than forty species of planktonic foraminifera are known. Aspects of their biology beyond that discussed in Chapters 3 and 6 are found in Anderson et al. (1979) and in Bé et al. (1977).

**The role of protozoa in plankton**   The previous discussion shows that nanoplankton protozoa are primarily responsible for the consumption of suspended bacteria and that nano- and microplankton protozoa consume a considerable fraction of the primary producers as well. As discussed in Section 7.2, the mineralization of nitrogen and phosphorus is coupled to carbon cycling and from the bioenergetic properties of protozoa we expect that about 30 percent of the ingested organic nitrogen (and phosphorus) will be mineralized by a protozoan feeding on food particles with a C:N ratio similar to its own. In addition, as much as 30 percent may be excreted in the form of particulate or dissolved organic material. Experimental work with pure cultures of heterotrophic microflagellates generally supports this prediction (Goldman et al., 1985; Sherr et al., 1983). The question of which organisms are responsible for nutrient regeneration in the plankton is of interest to marine ecology in its own right. In addition, direct measurements of nutrient regeneration by various functional groups of the plankton yield information on their role in the food chains.

Such experiments are carried out by fractionating plankton organisms into different size classes and using a $^{15}$N-isotope-dilution technique which allows simultaneous estimation both of the uptake and the remineralization of ammonia. Harrison (1978) found that in coastal waters some 39 percent of the nitrogen remineralization was due to the <1 μm (bacterial) fraction while the 1 to 35 μm fraction (presumably protozoa) was responsible for about 50 percent. For oceanic water samples Glibert (1982) found that most of the mineralization was due to organisms smaller than 10 μm. For a number of localities in coastal waters, the bulk of the mineralization was always due to organisms <200 μm, although the relative importance of different-size groups varied. Paasche & Kristiansen (1982) found that most of the mineralization in the Oslo Fjord was due to organisms within the size range of 45 to 200 μm. Since planktonic food chains, particularly those in coastal and more or less eutrophic waters, are never in a steady state it is not surprising that the relative contribution from each different size class varies over time. The observations do show that a very large or even a dominant part of the carbon flow of planktonic food chains must be channeled through protozoan populations. Since as much as 80 percent of the nitrogen is remineralized by protozoa, this means that they represent at least two trophic levels in the food chain (cf. Section 7.2).

Protozoa must represent an important food resource for metazoan plankters. Protozoa are within the size range of particles which many planktonic organisms, such as copepods, retain and it is safe to assume that the importance of protozoa as food for metazoa is proportional to protozoan abundance. However, there is still little direct evidence for this, perhaps because most protozoa, in contrast to armored dinoflagellates and diatoms, for example, do not leave identifiable remains in the guts or fecal pellets of animals. Thus the role of tintinnids as food for other organisms is best documented; their loricae often fill the intestine of planktonic invertebrates (Capriuolo & Ninnivaggi, 1982). Direct experimental studies are still scarce: Berk et al. (1977) showed that the planktonic copepod *Eurytemora* filters and eats suspended cells of the ciliate *Uronema*, and Pavlovskaya & Pechen (1971) demonstrated the ingestion of ciliates by copepods and larval barnacles, but concluded that "infusorian" food could not totally satisfy the nutritional demands of the studied species.

## 8.3    Marine sediments

**General biological properties of sediments**    By far the largest part of the bottom of the sea is covered by sediments. The surface layer of such sediments harbors a much higher concentration of life than does the water column above it. There are several reasons for this phenomenon. Interfaces between water and inert solids (and the air-water interface as well) tend to concentrate organisms. Thus bacteria and some microalgae adhere to surfaces and attract protozoa which are adapted to feed on them. Many suspension feeders attach (either temporarily or permanently) to solid surfaces although their food particles derive from the surrounding water. These protozoa also support populations of predators. However, more important than these purely mechanical effects is the fact that dead, particulate organic material (remains of plankton organisms, dead seaweeds, and benthic animals) accumulate in sediments due to the effect of gravity. The bacterial production supported by this material is the basis for phagotrophic food chains. Due to the degradation of organic material, the sediment is enriched in mineral nutrients relative to the water column. In the surface layers of shallow water sediments, dense populations of cyanobacteria and eukaryotic, photosynthetic organisms take advantage of this and they also support populations of phagotrophs.

Since the input of organic material is high relative to the supply of oxygen, which depends mainly on molecular diffusion (bioturbation of benthic animals and microalgal photosynthesis in the upper millimeters of shallow water sediments also play a role), sediments will be anoxic at a certain depth beneath the surface. The exact depth at which the environment becomes anoxic depends on several factors: the influx of organic matter, the mechanical properties of the sediment, water turbulence, and whether or not the surface is exposed to light.

Within the anoxic zone, organic material is degraded by anaerobic bacteria. The initial degradation is due to fermenting bacteria which generate hydrogen and low molecular weight organic compounds, such as acetate, propionate, and butyrate. These compounds are further mineralized by bacteria which employ respiratory processes with terminal electron acceptors other than oxygen. In marine environments, sulfate-reducing bacteria dominate because of the high concentration of sulfate in seawater. Thus in investigated cases, around half of the entire mineralization in sediments is channeled through sulfate reduction (Sørensen et al., 1979). The end product of this process is sulfate (at the prevailing values of pH, found mainly in the form of the $HS^-$ ions). The sulfide is responsible for the smell of anoxic sediments and the black color is due to the formation of ferrous sulfide. As the sulfide diffuses upward along the concentration gradient it will come in contact with oxygen. In this transition zone there are a large number of "chemolithotrophic" bacteria which make their living by oxidizing sulfide to elemental sulfur or further to sulfate. These organisms represent food for high numbers of phagotrophic organisms, especially protozoa.

In shallow waters where the sediment surface is exposed to light, further complexities arise. Light penetrates only a few millimeters beneath the sediment surface. But if the anoxic zone extends up to the photic zone of the sediment, photosynthetic bacteria (in particular green and purple sulfur bacteria) will thrive; these organisms utilize hydrogen sulfide as an electron donor in a photosynthetic process. Above this zone, cyanobacteria and photosynthetic eukaryotes (in particular diatoms, dinoflagellates, and cryptomonads) are found. Since all these photosynthetic organisms are only active in the light and since their activity tends to oxidize the environment, diurnal vertical migrations of the chemically defined zones (and thus of the microbes which belong to them) occur. Examples of such vertical zonations of chemical species, the effect of light and the zonation of microbial processes of some sediments are shown in Figure 8.5.

Where large amounts of organic material accumulate, the entire zonation may be compressed into a band which is only a few millimeters deep at the very surface of the sediment. If the water above the sediment is stagnant the oxic-anoxic transition zone may even extend above the sediment and into the water column. This phenomenon is most common in thermally stratified lakes, but it also occurs in sheltered and eutrophic marine environments and in deep basins and fjords where the water masses are thermally stratified.

While the vertical zonation of sediments is most obvious, horizontal patterns occur as well. Thus the zonation patterns described above also develop around large organic particles, such as dead animals or macroalgae which are buried in the sediment, and burrows of polychaetes and crustaceans which penetrate into anoxic sediments may be surrounded by an oxidized zone.

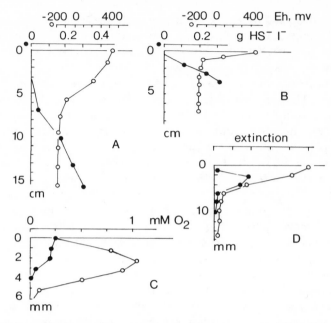

**Figure 8.5** Vertical zonation patterns in sediments. A and B. The vertical distribution of the oxidation-reduction potential and of free sulfide in a marine sand and in an estuarine sediment, respectively. C. The distribution of oxygen in the upper millimeters of an estuarine sediment in the dark (filled circles) and after two hours of exposure to light (open circles). D. The vertical distribution of chlorophyll *a* (open circles) and of bacteriochlorophyll *a* (filled circles) in an estuarine sediment (quantified as a function of the extinction coefficient at appropriate wavelengths of an ether extract of a standardized volume of sediment). (A,B: after Fenchel, 1969; B: after Sørensen et al., 1979; D: after Fenchel & Straarup, 1971.)

All these processes, as well as the spatial and temporal patterns which arise from them, allow for the great diversity of protozoa associated with sediments. In order to understand the functional properties of protozoan communities of sediments it is necessary to use sampling methods having very fine spatial resolution.

The mechanical properties of sediments are an important factor for the qualitative and quantitative distribution of protozoa. In well-sorted sands with small amounts of silt and clay, the space between the individual mineral grains harbors protozoa and other organisms. In sediments with a smaller average grain size, only smaller protozoa are capable of moving in the interstitial space, while in very fine-grained or poorly sorted sediments, protozoa (except for some foraminifera) are confined to the loose surface layer of flocculent material, since they are incapable of burrowing. Algal mats, detrital material, and masses of sulfur bacteria associated with the surface of aquatic sediments also harbor protozoan communities. The mechanical structure of such habitats is often loose and the interstitial space is large. These communities therefore include forms which are more or less planktonic and others which are permanently attached to algal filaments (a life form otherwise absent in sediments). Nevertheless, I have found it most natural to discuss these communities in conjunction with marine sediments.

**The interstitial fauna of sands**     During the 1930's, it was recognized that the interstitial spaces of well-sorted sands harbor a rich fauna of small animals adapted for movement between the sand grains of shallow water sediments and in the coastal ground water of beaches. This fauna was termed the "interstitial fauna" or "mesopsammon." Initially most emphasis was on metazoan microfauna and a wide variety of new and often unusual species was discovered. Kahl (1933) described a number of new species of ciliates from sandy sediments and Fauré-Fremiet (1950b) also contributed to the investigation of this fauna. Since then, a great variety of ciliates, some of which are quite unique, have been found in sandy sediments from a variety of coastal areas (e.g., Agamaliev, 1967; Burkovsky, 1970; Dragesco, 1960; Hartwig, 1973; a bibliography of interstitial ciliates was published by Hartwig, 1980); the ecology and quantitative significance of these organisms have been studied especially by Fenchel (1968, 1969) and by Hartwig (1973). Instial ciliates can be extracted quantitatively from samples of sand. The sand is placed in tubes closed below by nylon net which does not allow the sand grains to pass and the sample is covered by ice made from seawater. This creates a salinity gradient and a vertical flow of water which drives the organisms out of the sample (Fenchel, 1969).

The distribution and the quantitative importance of interstitial ciliates depend strongly on the size of the instial spaces. Since this is not easy to measure or calculate directly, the median grain size (of reasonably well-sorted sands) is usually taken as a measure of interstitial dimensions. Figure 8.6 shows the quantitative distribution of ciliates and their importance relative to the metazoan interstitial forms found in some sandy, shallow water localities along the Danish coast. It is seen that sands with a median grain size of between 0.12 and 0.25 mm harbor the richest faunas of ciliates. In finer sediments the interstitial spaces are too small to allow ciliate movement and burrowing microfauna such as nematodes dominate, while most protozoa are confined to the sediment surface. In coarser sediments, metazoan interstitial organisms, such as turbellarians, gastrotrichs, harpacticoid copepods, and archannelids have largely replaced ciliates in the niches.

The most obvious characteristic of interstitial ciliates (which they share to a large extent with the metazoan mesopsammon) is their mor-

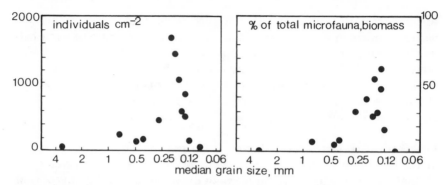

**Figure 8.6** (Left) numbers of interstitial ciliates and (right) their biomass as a percentage of ciliates + interstitial metazoa, in marine sediments in Danish waters both expressed as functions of median-sediment particle size (increasing from right to left). (After Fenchel, 1969.)

phological adaptation to interstitial life. The larger forms (some of these ciliates are extraordinarily long, often a millimeter or more) are oblong or leaf-shaped. Often they are ciliated only on one side, on which they slide along the surfaces of the mineral grains. Most forms are capable of adhering temporarily with thigmotactic cilia. In Figure 8.7 some characteristic interstitial ciliates are shown. Most, but not all of these belong to the ciliate order Karyorelectida; common genera are *Tracheloraphis, Remanella* (the species of which are often numerically dominant, see also Figure 7.1), *Kentrophorus,* and *Geleia.* Extremely oblong or flattened forms are also found among the prostomatids, the haptorids, and the heterotrichs (e.g., *Pseudoprorodon, Loxophyllum,* and *Blepharisma*). Smaller ciliates show less extreme morphological adaptations to this environment; common forms belong to the prostomatids (e.g., *Coleps*), the scuticociliates (e.g., *Cristigera* and *Pleuronema*), the cyrtophorids (e.g., *Chlamydodon*), and the hypotrichs (e.g., *Discocephalus, Aspidisca,* and *Diophrys*). The oligotrich ciliates, which otherwise are mainly planktonic, also occur in the interstitial environment. They are represented by species belonging to the genus *Strombidium,* but the interstitial species are flattened and capable of adhering to solid surfaces.

Although ciliates are the dominant and most obvious protists of sandy sediments, other groups are represented as well, but they have not received nearly as much attention. Phagotrophic flagellates (in particular, dinoflagellates such as the genus *Amphidinium*), euglenids, and bodonids are common (Dragesco, 1965; Fenchel, 1969), and naked as well as testacean amoebae occur (see Golemansky, 1978; Golemansky & Ogden, 1980; Page, 1974; Sudzuki, 1979).

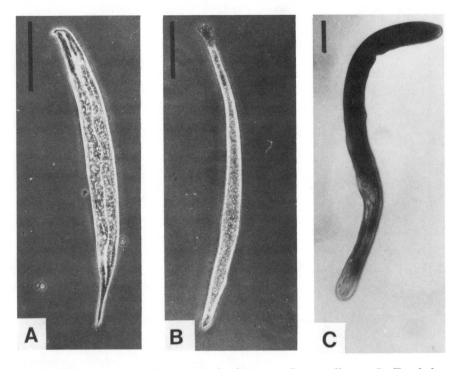

**Figure 8.7** Three typical interstitial ciliates. A: *Remanella* sp., B: *Tracheloraphis* sp., C: *Pseudoprorodon arenicola* from a sandy sediment in Limfjorden, Denmark. Scale bars: 100 μm.

The protozoan fauna of sands shows characteristic vertical zonation patterns in accordance with the chemical and microbial zonation (Figure 8.8). Three fairly distinct ciliate communities can be recognized. Their vertical extent depends on the depth of the oxic-anoxic boundary layer and on the steepness of the oxygen gradient and thus varies on a seasonal and a diurnal basis. The oxidized surface layer includes many scuticociliates, haptorids, cyrtophorids, oligotrichs, and hypotrichs, as well as genera like *Tracheloraphis* and *Frontonia*. Around the oxic-anoxic boundary layer a quite different ciliate fauna is developed. It is often dominated by species of *Remanella*, but includes a variety of other forms such as *Kentrophorus*, and also representatives of the above-mentioned groups and of the heterotrichs. The deepest zone is entirely anoxic and comprises a fauna of anaerobic ciliates. The most important ones are heterotrichs (*Metopus, Caenomorpha, Parablepharisma*), the related odontostomatids (*Saprodinium, Myelostoma*), and trichostomatids (*Sonderia, Plagiopyla*).

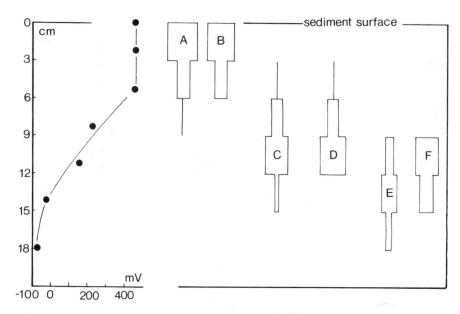

**Figure 8.8** (Left) the vertical oxidation-reduction profile and (right) the distribution of six species of ciliates in a sandy sediment. A: *Pleuronema coronatum*, B: *Frontonia arenaria*, C: *Remanella margaritifera*, D: *R. brunnea*, E: *Caenomorpha levandri*, F: *Metopus contortus*. The entire sample (a core sample representing 1 cm$^2$ surface) contained 1200 ciliates belonging to 32 species as well as some unidentified ones. (Data from Fenchel, 1969.)

The food of the fauna of interstitial ciliates consists of various types of microalgae (including diatoms, which, in spite of aphotic conditions, often occur in relatively large numbers many centimeters below the sediment surface), flagellates, other ciliates and bacteria (including cyanobacteria and large sulfur bacteria). The composition of interstitial ciliate fauna depends on the relative availability of food items. During the summer and in localities with a rich development of microalgae in the surface layers, species feeding on these organisms dominate; otherwise bacterivorous ciliates are usually the most common.

The overall role of protozoa in the carbon flow of marine, sandy sediments is not clear and quantitative estimates of their abundance have so far been restricted to ciliates. Based on the quantification of ciliates (and of metazoan microfauna) in a variety of marine sediments by Fenchel (1969) there can be little doubt that in certain types of sands (see Figure 8.6) ciliates dominate as consumers of microalgae and as consumers of bacteria (possibly together with other types of protozoa). In other types of sediments they share this role with micrometazoa, and in fine sediments they also share with detritus feeders, which more or less indiscriminately ingest the bulk sediment. Biotic communities within sediments characteristically do not share the simple size structure and the relationship between size and position within phagotrophic food chains, which is typical of the plankton. This is because life within sediments is constrained by mechanical properties which may exclude certain size classes of organisms. In sediments consisting of clay and silt, bacteria are probably the only truly interstitial organisms. These are exploited by organisms which are sufficiently large and robust to burrow. In such cases ciliated protozoa and flagellates may play a modest quantitative role.

**Silt and clay**   Well-sorted, sandy sediments constitute only a narrow zone along marine coasts. In most areas, at depths of 15 to 20 meters or greater, the sediments contain so many fine particles that an interstitial fauna does not occur. By far the greatest part of shelf sediments consist of silt and clay. In such sediments, ciliates, flagellates, and amoebae are confined to the flocculent surface layer. This fauna is very poorly studied. This is in part due to methodological problems since (in contrast to the interstitial fauna which can be extracted quantitatively) the enumeration of protozoa of detrital or flocculent sediments depends on direct sorting or dilution cultures, methods which are often unreliable and tedious.

A few samples of the upper millimeters of such sediments studied by Fenchel (1969) suggest that numbers of ciliates are ten to one hundredtimes lower (on an area basis) than in sandy sediments—less than 100 cells/cm$^2$. This fauna seems to be dominated by hypotrichs and scuticociliates. Lighthart (1969) and Mare (1942) attempted to quantify protozoa using dilution-culture techniques and demonstrated the presence of ciliates and amoebae, as well as heterotrophic flagellates, in the surface layers of off-shore sediments. Lighthart (1969) demonstrated a high diversity of microflagellates, mainly bodonids and nonpigmented chrysomonads, in the surface of sediments ranging from about 20 to 1500 meters deep. Typically, around 10$^3$ flagellates per ml of sediment was found. The quantification method, as well as the fact that it is difficult to isolate the sediment layer which contains protozoa from the bulk of the core sample, suggest that these figures underestimate the abundance of flagellates; even so, the results suggest that microflagellates play a considerable role as grazers of bacteria in the surface layers of sediments.

Among the protozoa of noncapillary sediments the foraminifera have enjoyed most attention. These organisms are found on or in sediments and on solid surfaces at all depths, but their numbers and diversity seem to be greatest in off-shore sediments. While species with calcareous tests dominate in shallow waters, forms which build tests from mineral grains,

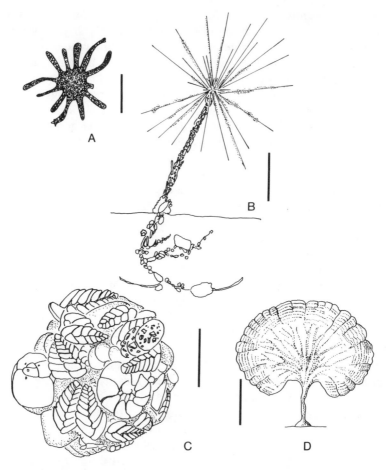

**Figure 8.9**  Shelf and deep-sea sarcodines. A-C: foraminifera; D: a xenophy-ophorian. A. *Astrorhiza limicola*, common in, e.g., North Sea sediments (scale bar: 5 mm). B. *Marsipella arenaria* found in the Oslo Fjord at 60–150 m depth (scale bar: 5 mm). C. *Psammosphaera testacea* which builds its test from shells of other foraminifera; found in the Oslo Fjord at 100–200 m depth (scale bar: 1 mm). D. *Stannophyllum flustraceum* collected in the Central Pacific at 5000 m depth (scale bar: 5 cm). (A: redrawn from Gabel, 1971; B,C: redrawn from Christiansen, 1958; D: redrawn from Tendal, 1972.)

or sponge spicules, etc. predominate in deeper waters. Also a number of strange and large types of foraminifera appear at greater depths (Figure 8.9). Some of these species burrow in the sediment and are found several centimeters beneath the surface. *Allogromia*, which resembles and probably often is mistaken for fecal pellets of invertebrates, holds its test perpendicular to the sediment surface while its pseudopodia penetrate into the underlying sediment; according to Nyholm & Gertz (1973) these organisms are so numerous that they stabilize the detritus layer of soft sediments.

Systematic accounts of the foraminiferan fauna of northern European shelf sediments are found in Christiansen (1958), Gabel (1971), and Höglund (1947). These papers disclose a surprising diversity of forms which must reflect a corresponding niche diversification. As regards prey capture, some results are given by Christiansen (1958, 1964, 1971). Food items of the benthic foraminifera include various protozoa (including other foraminifera) and bacteria.

The quantitative importance of foraminifera increases with increasing depth, both relative to small invertebrates of comparable size ("meiofauna": nematodes, harpacticoids, etc.) and in absolute terms. Thus, in the North Sea generally fewer than 100 individuals per g of sediment are found in the shallower parts, while in the deeper parts of the Skagerak Sea and in the North Atlantic Sea, this number may exceed 1000 individuals per g (Gabel, 1971). Gerlach et al. (1985) studied the quantitative composition of the fauna in soft sediments at a location in the North Sea. They found about 5000 foraminiferan cells/cm² which formed a biomass exceeding that of all the other benthic animals together. Although the authors admit that they may in some cases have been unable to distinguish living foraminifera from empty tests, their results document the importance of these organisms in soft, marine sediments.

Although the evidence is of a qualitative nature, the presence of foraminiferan tests in the guts of various benthic invertebrates illustrates one role of these organisms in benthic food chains. Foraminifera are eaten by many species of prosobranchs, bivalves, polychaetes, and other invertebrates; some nematodes seem to have specialized on this diet (Lipps & Valentine, 1970; Sliter, 1971).

**Deep sea sediments**    Although the protozoan fauna of sediments of the shelf may be relatively unknown, this applies even more so to the sediments of the deep sea. For ciliates and flagellates, apparently the only investigations are those of Burnett (1973, 1977, 1981, and references therein). He was able to demonstrate the presence of these groups, as well as of amoebae and some testate cells, and to offer some estimates of their numbers based on fixed core samples. Protozoan biovolume in the upper 5 mm of the sediment corresponded to about 0.2 mg wet weight per g sediment. Flagellates were quantitatively dominant.

In addition to these "normal" protozoa, the deep sea harbors some unusual sarcodines, the Xenophyophoria. These are giant, plasmodial forms of a characteristic shape, often several centimeters long (Figure 8.9,D). They contain crystals of barite and foreign particles (shells of radiolaria and foraminifera, sponge spicules, and mineral grains) which together make up the bulk of the organisms. They occur on the surface of deep-sea sediments in all oceans, but they seem to be most diverse in the Pacific; there are also a few records of xenophyophorians from more shallow waters. Although they have been collected by deep-sea expeditions since the Challenger Expedition, very little is known about their biology. They produce gametes as a part of their life cycle and for that reason they are usually classified with the Reticulopodia. They are assumed to feed on bacteria which are digested outside the test since accumulations of food particles are found on the surface of preserved specimens. The pseudopodia have never been observed. Tendal (1972) has written a monograph on the subject of the xenophyophoria.

The deep sea probably harbors other "odd" protozoa: Large nontestate sarcodines have occasionally been observed (popularized by Haeckel's "Urschleim" theory for the origin of life), but some of these findings have later been disclaimed as artifacts of preserved sediment samples. However, there is no doubt that such large naked sarcodines do exist. Nyholm (1950) observed living specimens of an amoeboid organism, *Megamoebomyxa*, which is found in muddy sediments at a depth of only about 70 m in the Gullmarfjord on the west coast of Sweden. This creature of unknown taxonomic affinity measures 2 to 2.5 cm in length.

**Algal mats, tidal pools, and sulfureta**    These types of habitats develop on the surface of sediments in shallow and sheltered localities, such as estuaries, shallow water bays, coves, and pools. Their physical structure is due to the presence of various types of filamentous algae and prokaryotes (*Oscillatoria*, *Beggiatoa*, etc.). Among these filaments there is a luxuriant growth of unicellular, photosynthetic eukaryotes, in particular diatoms, but also dinoflagellates, euglenids, and crypto-monads. In contrast to pelagic habitats and to the interstitial spaces of sands, these habitats do not favor any one particular type of motility or morphological shape. They harbor forms which lead a planktonic life within large interstitial spaces and forms which slide or walk along surfaces, as well as forms permanently attached to algal filaments.

Different types of shallow water biota have many protozoan species in common, but it is still possible to demonstrate some characteristic protozoan faunal assemblages. Many shallow water localities are strongly influenced by freshwater intrusions. Although their protozoan faunas are basically marine in character until very low salinities (1–3 per mille S) are reached (ordinary seawater has a salinity of 30–35 per mille), some effects of salinity are evident. Thus the diversity of foraminifera is very low in localities exposed to constant, dilute brackish water or to strongly fluctuating salinities, while at least one group of ciliates (the karyorelictids) are so stenohaline that they do not occur at salinities below about 10 per mille. On the other hand, at salinities of around 7–10 per mille, some euryhaline, freshwater ciliates appear (e.g., species of *Spirostomum, Climacostumum*, and *Stentor*). The most important distinction between these habitats, however, is between those which are strongly influenced by anaerobic conditions and hence by bacteria associated with the microbial sulfur cycle ("sulfureta") and between those which are permanently aerobic. Intermediate situations do, of course, exist.

The protozoan faunas of salt marshes and algal mats in shallow waters have often been described, in particular with respect to ciliated protozoa and to foraminifera (e.g., Borror, 1972; Dietz, 1964), while Fenchel (1969), Lee (1980), Lee & Muller (1973), Matera & Lee (1972), and Webb (1956) include ecological aspects. The ciliate fauna of such localities is rich in species and this reflects the diversity of available food particles and of feeding mechanisms. Microorganisms attached to solid surfaces are in particular exploited by cyrtophorid ciliates; diatoms by species of *Chlamydodon*, filamentous prokaryotes mainly by various dysterids (see Figure 3.7). Many suspension-feeding ciliates are also associated with solid surfaces: Scuticociliates such as *Cyclidium* (Figure 3.4,C) and *Pleuronema* feed on small particles, mainly bacteria; while a variety of hypotrich and heterotrich ciliates (including species of *Euplotes, Diophrys, Holosticha, Condylostoma*, and *Blepharisma*) feed on larger particles, such as microalgae and flagellates. Also predatory ciliates (*Litonotus, Loxophyllum*) slide along surfaces and feed mainly on other ciliates. Free-swimming species of *Frontonia* (see Figure 3.8,B) feed on large diatoms and filaments of *Oscillatoria*, while holophryids, such as species of *Prorodon, Coleps*, and *Lacrymaria*, are predators on flagellates and other ciliates. Some *Coleps* and *Prorodon* species also thrive as histophages. Species of *Strombidium* (Figure 3.8,C) live a planktonic life among algal filaments, while on the other hand various kinds of peritrich ciliates and the predatory suctorians (Figure 3.8,D) are permanently attached to larger algal filaments. The fauna of heterotrophic flagellates is less well documented, but colorless euglenids, chrysomonads, bodonids, and the permanently attached choanoflagellates abound. Except for the most brackish localities, foraminifera are also common, either attached to larger algae or on the surface of the underlying sediment. The most frequent forms are species belonging to the genera *Elphidium* (Figure 3.10), *Ammonia, Miliammina*, and *Trochammina*. The ecology of giant foraminifera in tropical, shallow water habitats is discussed in Section 6.2.

Sulfureta develop where large amounts of organic material, such as leaves of seagrasses or thalli of large algae, accumulate in the sediment or on its surface in relatively sheltered areas. Here the key process is microbial sulfate reduction which leads to a copious production of sulfide. On the surface of the sediment or in the detrital accumulations dense patches of either "white" or "purple sulfur bacteria" occur. The former oxidize sulfide with oxygen; they form white patches due to the intra- or extracellular accumulation of elemental sulfur. At times the purple sulfur bacteria dominate and color large areas of the bottom; they are photosynthetic anaerobic organisms which utilize sulfide as an electron donor.

In temperate regions the dominant physiological type of bacteria depends on the season; from autumn to early spring when light conditions are poor, flocculent layers of the filamentous, white sulfur bacterium, *Beggiatoa*, predominate at the surface, while during the summer the same patches are purple due to dense accumulations of photosynthetic bacteria. As a result of both anoxygenic and oxygenic photosynthesis, sulfureta show strong diurnal changes in the chemistry of the surface layers followed by migrations of the microbial populations. Thus, during the night when the top surface layer and even the water above it, becomes anoxic and contains hydrogen sulfide, populations of white sulfur bacteria migrate upward and cover the photosynthetic organisms and the sediment surface with a white veil. As the sun rises they disappear beneath the surface again.

The microbial life of sulfureta is rich and diverse. In addition to the different types of chemolithotrophic and photosynthetic bacteria, a variety of cyanobacteria, in particular the filamentous *Oscillatoria*, are common, as well as many eukaryotic photosynthetic organisms, in particular euglenids and dinoflagellates. The ecology of sulfureta has been studied by Fenchel (1969). The surface landscapes of two different sulfureta (one dominated by *Beggiatoa* and the other by the purple sulfur bacterium *Chromatium*) are shown in Figures 8.10 and 8.11.

The ciliate fauna of sulfureta is very rich both qualitatively and quantitatively: A sulfuretum in a shallow bay contained nearly fifty species of ciliates and in the surface layer alone there were more than a thousand cells per ml. The different species undoubtedly show vertical zonation patterns as discussed for sand ciliates (compare Figure 8.8), but the chemical gradients of sulfureta are often so steep that it is difficult to sample sufficiently precisely to demonstrate this pattern. Different types of sulfur bacteria constitute a dominant food source for many of the resident ciliates; others depend on cyanobacteria, diatoms, flagellates, and other ciliates.

Among the ciliate fauna of sulfureta, the anaerobic species represent a special group. The most common genera are *Metopus, Caenomorpha, Parablepharisma, Saprodinium, Myelostoma, Plagiopyla*, and *Sonderia*. Species belonging to the last genus are specialized to feed on large, filamentous bacteria (Figure 3.8,A). Many other ciliates which do not belong to this specialized group of anaerobes also occur: Among

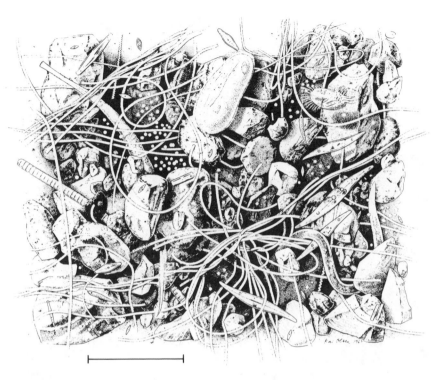

**Figure 8.10**   The microbial community on the surface of an estuarine sediment dominated by the filamentous sulfur bacterium, *Beggiatoa*. In addition to different ciliates (*Tracheloraphis, Frontonia, Diophrys, Trochiloides*) various diatoms, euglenids and a nematode worm are seen. Scale bar: 300 μm (Fenchel, 1969).

them, species of *Blepharisma, Gruberia, Trochilia*, and the marine *Paramecium calkinsi* feed mainly or exclusively on sulfur bacteria; scuticociliates (*Cyclidium, Cristigera*) feed on smaller bacteria; *Frontonia, Peritromus*, and *Chlamydodon* may be the most common algivores; while *Prorodon* and *Paraspathidium* are the principal predators of protozoa.

Sulfureta are not found exclusively in shallow waters; in eutrophic coastal waters, white patches of *Beggiatoa* or other white sulfur bacteria are often seen at considerable depths. Their protozoan faunas have not been investigated; it seems likely that they resemble those of shallow water sulfureta, but they are probably less diverse since photosynthetic organisms must be absent.

**Detrital material**   Particulate detrital material deriving from macrophytes, such as seagrasses, mangrove leaves, large algae, and also

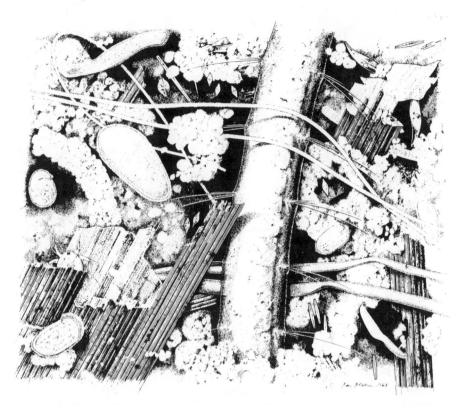

**Figure 8.11**  The microbial community at the surface of an accumulation of purple sulfur bacteria. In addition to ciliates (*Paraspathidium, Frontonia, Plagiopyla, Peritromus, Prorodon, Paramecium calkinsi, Blepharisma*), diatoms, a harpacticoid nauplius larva, and an oligochaete worm transversing the scene, lumps of sulfur bacteria and decaying eelgrass leaves as seen. Scale as in Figure 8.10 (Fenchel, 1969).

dead phytoplankton cells, are found in marine habitats, most frequently, of course, on shallow water sediments. These particles may also be transported far off shore by tides and currents and they may be found on the sediments beyond the shelf. Detrital particles may also be held in suspension in turbulent water. Characteristic microbial communities develop on them; these are easily studied in the laboratory and they illustrate some general features of microbial phagotrophic food chains. Quite similar communities, with quite similar properties, develop on detrital particles in freshwater.

If previously sterilized particles (typically around one millimeter in diameter), consisting of tissue from vascular plants, are placed in seawater a characteristic succession of microorganisms can be seen during

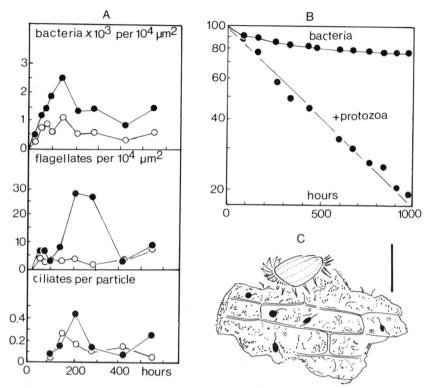

**Figure 8.12** A. Successions of bacteria, flagellates, and ciliates on water-extracted hay particles in seawater; filled circles represent an experiment enriched with nitrate and phosphate. B. Decomposition rate (as percent organic carbon remaining) of barley hay homogenously $^{14}$C-labelled in seawater with bacteria and with bacteria plus protozoa. C. A microbial community on a detrital particle (scale bar: 50 μm). (A,B: redrawn from Fenchel & Harrison, 1976.)

the following days (Fenchel, 1970; Fenchel & Harrison, 1976; Stuart et al., 1981; also Figure 8.12,A,C). During the first 50 hours or so, bacteria colonize the surfaces of the detrital particles to reach typical densities of 1 to 2 × $10^4$ cells per $10^4$ μm$^2$, or somewhat less than 10 percent of the surface. At this time populations of heterotrophic flagellates appear, consisting mainly of bodonids, colorless euglenids, and chrysomonads. These populations peak after another 50 to 100 hours during which time the bacterial numbers decline. After several hundred hours, ciliates (mainly small scuticociliates like *Uronema* and *Cyclidium*, but also hypotrichs and other forms) appear. After about 200 to 400 hours these populations remain relatively stable as long as the detrital particles maintain their integrity. Other groups of organisms, including fungi and nematodes, may eventually develop on or in the detrital particles. The bacterial populations and the protozoan populations each have about the same biomass. Since the microbial communities develop mainly on the surfaces of the particles their absolute numbers per unit weight of particles depend on particle size.

These successional events are likely to occur repeatedly on such particles in nature. Fenchel (1970) studied microbial communities on particles of seagrasses and mangrove leaves in a shallow cove on the coast in Florida. Within the accumulations of detritus, amphipod crustaceans fed on the particles. Although they ingested the particles, they apparently only digested the microbial communities since the detrital particles reappeared in fecal pellets. These pellets were again colonized by bacteria and protozoa, and eaten again by the amphipods. The main effect of the amphipods on the detrital particles was to diminish the average size of a particle on each passage through the digestive system, thus increasing the microbial population per unit volume of detritus and accelerating their microbial degradation.

A question which has attracted some attention is the effect of protozoan grazing on the primary decomposers—the bacteria—and on the rate of degradation of detrital material. Empirical evidence shows that the decomposition of particulate detritus is accelerated in the presence of protozoan grazers, as compared to a situation in which only bacteria are present (e.g., Fenchel & Harrison, 1976; Sherr et al., 1982, and references therein; Figure 8.12,B). In detrital material containing no protozoa, the density of bacteria is on the average about twice that of material with protozoa present. The general reason for the stimulation of bacterial activity when they are grazed must be that in the absence of grazing, the bacterial populations are controlled by some limiting resource so that when they reach their maximum size, growth of the individual cells is very slow. Since the work by Johannes (1965), this effect of grazing has generally been attributed to the remineralization by protozoa of nutrients otherwise bound up in the bacterial biomass.

It can be demonstrated that the degradation of detrital material (which consists mainly of structural carbohydrates of plants) in sea or freshwater is often limited by phosphorus and especially nitrogen (see Figure 8.12,A) so that the addition of such mineral nutrients increases microbial population sizes as well as the decomposition rate. Thus protozoan mineralization of ingested bacterial cells should, according to this view, maintain a somewhat higher concentration of ammonia and orthophosphate in the water of the experimental containers, thus allowing for a higher bacterial activity. More or less direct evidence for this is found in Sherr et al. (1982) and a review and general discussion is found in G.T. Taylor (1982). Various experiments aimed at demonstrating the effect of protozoa on mineral regeneration have given somewhat different and ambiguous results, which have not always been rigorously analyzed. It would seem that this effect will play only a small role if the total initial amounts of nutrients are very low or very high, but at an intermediate concentration of dissolved minerals the protozoa may accelerate the decomposition rate through the recycling of nutrients. However, none of these experiments can readily be extrapolated to nature, where detrital decomposition does not take place in closed systems with a finite access to mineral nutrients.

**Solid surfaces** Many ciliates and other protozoa are associated with the surfaces of solid objects, such as sheltered rocks, the thalli of some species of algae (others seem to be permanently free of protozoa and any other kind of epifauna), and artificial submerged substrates, but little is known of the ecological structure of these communities. Characteristic species in these communities are permanently attached forms, notably the peritrichs (species of the solitary *Vorticella*, the colonial *Zoothamnium*, and the loricate *Cothurnia*). Unique to marine (and brackish) environments are the heterotrich, folliculinid ciliates. These are brightly colored ciliates which build ampulla-shaped loricae. Both these groups are suspension feeders deriving their food particles from the surrounding water. The suctorians (Figure 3.8,D) are carnivorous and feed on the motile ciliates of the habitat. Dysterid ciliates (*Dysteria, Hartmannula, Trochilia*) feeding on cyanobacteria and filamentous bacteria are also common, as are hypotrich ciliates (such as *Euplotes, Aspidisca, Holosticha, Keronopsis*, etc.). Amphileptids (*Litonotus, Hemiophrys*) seem to be the dominant carnivores; some of them are specialized to prey on the zooids of peritrich ciliates.

Two studies (Agamaliev, 1974, in the Caspian Sea, and Persoone, 1968, in the polluted harbor of Ostende, Belgium) followed the colonization of ciliates on submerged slides and studied the composition of their associated ciliate faunas through the year. Agamaliev recorded a total of 130 species of ciliates while about 40 species were found by Persoone.

In addition to ciliates, a variety of other types of protozoa are found associated with surfaces in the sea. These include euglenid and bodonid flagellates, small amoebae, foraminifera, and heliozoa. Apparently, no ecological studies on this fauna have as yet been published.

# 9

# Protozoan Communities: Freshwater Habitats

## 9.1 Differences from marine communities

Comparisons between marine and limnic (or freshwater) faunas show large differences with respect to the taxonomic groups represented. Also when related forms are found they often fill different ecological niches in the two types of environment. In the sea, insects, leeches, pulmonate snails, and cladocerans are either completely absent or quantitatively unimportant, filling only specialized or marginal niches. Conversely, corals, tunicates, echinoderms, and other important groups of invertebrates are entirely absent from freshwater. With regard to protozoan communities, differences between limnic and marine habitats are less dramatic.

Two major groups of protozoa do not occur in freshwater. One of them, the radiolaria, has no ecological counterpart in freshwater. The foraminifera, however, are to some extent replaced by testaceans. Although testate amoebae do occur in the sea, their quantitative importance and diversity is much higher in freshwater (and terrestrial) environments.

The other major groups of protozoa do not show such differences between freshwater and marine environments. In most cases among the heterotrophic flagellates, for example, congeric (or sometimes possibly identical) species fill identical niches in freshwater and in the sea. Exceptions include species such as the acanthoecid choanoflagellates and the large phagotrophic dinoflagellates, both of which are absent in freshwater.

134

Comparisons between limnic and marine faunas of ciliates also show that congeneric or closely related forms tend to fill identical niches in comparable habitats. There are a few notable exceptions to this. The karyorelictid ciliates are represented in freshwater by only one genus, *Loxodes*. The modest representation of karyorelictids is related to the fact that a characteristic fauna of limnic interstitial ciliates is not found (cf., Section 8.3). The interstitial spaces of limnic, sandy sediments are populated by ciliate species which to a large extent also occur in other types of benthic communities. These ciliates do not show special morphological adaptations to interstitial life (see Dragesco, 1960). There are also a few examples of ciliate families which occur principally in either marine or in limnic environments. For instance, the sessile, folliculinid ciliates do not occur in freshwater and the colpodid and tetrahymenine ciliates, which in limnic environments exploit patchy mass occurrences of bacteria, do not play a similar role in the sea.

Heliozoa seem to be more important in freshwater environments than in the sea; this can perhaps (as in the case of testaceans) be related to the absence of foraminifera, but marine heliozoa have been rather neglected as objects of study so their importance may be underestimated. Naked amoebae are of similar importance and fill similar niches in the two types of aquatic environments, but tend to be represented by different species in each environment (Page, 1976, 1983).

A general feature of freshwater protozoan communities (as compared to marine ones) appears to be that species assemblages of various habitats are less distinct. This characteristic is not easy to quantify or document, but it seems obvious to protozoologists who are familiar with both freshwater and marine faunas. The absence of a characteristic interstitial fauna in freshwater was already mentioned. While typical planktonic ciliates occur in freshwater, the plankton often includes benthic forms as well. This in particular applies to monomictic lakes in which the oxycline may be totally dominated by benthic ciliates.

Many individual species of limnic protozoa seem to occur in a wide variety of habitats. Two reasons for this can be suggested. Lakes and ponds are relatively small and on a geological time scale they are ephemeral. This should lead to a lower rate of speciation and a higher rate of extinction, processes that are generally held to explain the relative paucity of limnic animal species. (This view is supported by the fact that the relatively rare, geologically old lakes do harbor diverse, endemic faunas. If the reports of a strange, endemic fauna of specialized plankton ciliates in Lake Baikal are true, see Section 7.5, this explanation may also apply to protozoa.) It is probably also significant that the different types of freshwater habitats are usually quite small as compared to the size of corresponding marine habitats. Many freshwater bodies are quite shallow. Also, areas of well-sorted sand are usually sparse and of restricted size in lakes. Such relatively small patches may not have allowed the evolution of freshwater habitat specialization to occur to the same extent as in the sea.

The above considerations suggest that the total number of species of freshwater protozoa is lower than that of marine ones. If the comparison includes radiolaria and foraminifera, this is almost certainly true. If these groups are excluded it may still be the case, but this is not really known.

## 9.2   Pelagic protozoa

As in the case of marine plankton, the ecological role of protozoa in limnic pelagial regions has only recently begun to be studied, although the taxonomy and the biology of individual forms has been studied for a long time. The reason for this is that protozoa are largely lost or rendered unrecognizable by the use of traditional methods to quantify plankton. It also appears that in freshwater, the rotifers occupy many ciliate niches. We should remember, however, that in general, metazoan plankters cannot account for the turnover of the smallest plankton organisms, in particular, bacteria (e.g., Pedrós-Alió & Brock, 1983).

The protozoan nanoplankton, represented mainly by the heterotrophic flagellates, has been ignored from an ecological viewpoint although a considerable amount of taxonomic work has been carried out. The view of Skuja (1956) may be mentioned as an example of earlier light-microscope studies and Fenchel (1986b) gives references to more recent studies on individual groups. Sorokin & Paveljeva (1972) were the first to quantify these organisms during a study of the pelagial food chains of a lake in Kamchatka. They showed how the number of heterotrophic flagellates increased following a bacterial bloom; indirect evidence suggested that these protozoa are the main consumers of bacteria in this particular lake. The peak concentration of flagellates in the water column was about $10^3$ cells/ml which is comparable to similar figures from the sea.

Sherr et al. (1982) studied the microbial succession which followed a bloom of dinoflagellates in Lake Kinneret in Israel. A substantial part of the dinoflagellate production was not grazed by zooplankton and the dead cells were degraded by bacteria. The resulting bloom of bacteria was in turn utilized by heterotrophic flagellates. Riemann (1985) studied several plankton components (including bacterial production) in a Danish lake and in experimental enclosures of lake water. He found that in the lake and in enclosures containing fish, grazing by heterotrophic flagellates could account for the turnover of the entire bacterial production. If fish were excluded, however, zooplankton invertebrates (in particular cladocerans) increased in numbers whereas flagellate numbers decreased and the latter could not fully account for the bacterial turnover. This effect may be due in part to cladoceran grazing of the flagellates and in part to direct grazing of cladocerans on bacteria (unfortunately, ciliates were not included in this study). Thus, this work does show that under normal conditions flagellates may be the main consumers of bacteria in lake water; it also suggests experimental approaches which may be useful for future studies on phagotrophic food chains.

On the basis of the limited evidence in the literature and on my own unpublished observations, numbers of heterotrophic flagellates in lake water are comparable to those found in the sea (i.e., typically around a thousand cells per ml) although large fluctuations in numbers are typical of eutrophic waters where numbers may be ten times higher following bacterial blooms. The composition of this flagellate fauna generally resembles its marine counterpart (except for the absence of acanthoecid choanoflagellates), being comprised mainly of choanoflagellates, colorless chrysomonads, and helioflagellates. Since the surface film (the neuston) and the suspended particles constitute relatively important patchy habitats in both shallow and eutrophic lakes, flagellates which are attached to or associated with surfaces, such as attached choanoflagellates, bicoecids, bodonids, and colorless euglenids also play significant roles (Figure 9.1,A). However, in contrast to most marine pelagic environments, bacterivorous ciliates are also important components of the plankton of lakes under some circumstances; this is supported by the observation that bacterial numbers may exceed $10^7$ cells/ml in eutrophic lakes, and bacterivorous ciliates seem only to be competitive at fairly high densities of bacteria (see Section 8.2).

The quantitative role of plankton ciliates is better documented than that of heterotrophic flagellates. The epilimnion of lakes harbors a considerable diversity of species. Oligotrich ciliates often dominate (this is almost always the case in the sea) and tintinnids, as well as naked forms, are present. The former are not nearly as diverse as they are in seawater and they are mainly represented by species belonging to the genera *Codonella, Tintinnopsis*, and *Tintinnidium* (Figure 9.1,B,C; for a taxonomic treatment of limnic tintinnids, see Foissner & Wilbert, 1979). The naked oligotrichs include *Halteria* (Figure 3.4,B) and species of *Strombidium*, among which *S. viride* contains zoochlorellae. Hecky & Kling (1981) found that in Lake Tanganyika the biomass of *Strombidium viride* equalled the phytoplankton biomass; thus it is likely that this ciliate may play a substantial role as a primary producer. Like their marine counterparts, limnic oligotrichs consume relatively large ($>2$ $\mu$m) particles, i.e., mainly flagellates and small phytoplankton cells. The ecological role of the prostomatid *Mesodinium* is not clear. The freshwater plankton also includes ciliates which are not typically pelagic

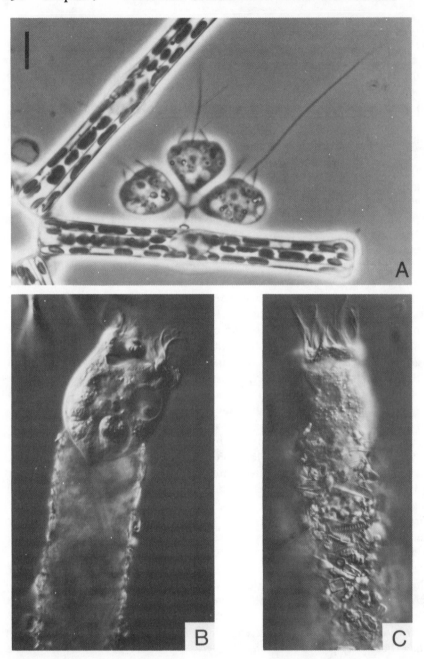

**Figure 9.1**    A. The choanoflagellate *Codosiga* attached to the planktonic diatom *Asterionella* (scale bar: 5 µm). B,C. The limnic tintinnid *Tintinnidium* sp. All from Esthwaite Water, English Lake District. (A: photograph by Hilda Canter-Lund; B,C: photographs by Bland J. Finlay.)

forms; among these, species of *Frontonia* and *Euplotes* (Figure 9.2,A) are common. The bacterivorous ciliates, which may show mass occurrence following bacterial blooms around the oxycline in monomictic or meromictic lakes, are also atypical plankton forms. Among them, scuticociliates (*Cyclidium*, see Figure 3.4,C, and *Uronema*) are the most important forms; also, stalked but unattached peritrichs, such as *Vorticella*, may occur in large numbers in plankton. The ciliate predators consist of prostomatids, such as *Didinium, Coleps, Acaryophrya*, and *Actinobolina* (see Figure 9.2,B,C) and pleurostomatids, such as *Paradileptus* and *Trachelius* (which feed on rotifers). Also the hymenostome *Lembadion*, which feeds on flagellates and on other ciliates, may occur in plankton (Figure 9.2,D).

The quantitative importance of ciliates was studied in a eutrophic lake by Pace (1982) and by Pace & Orcutt (1981). They found numbers ranging from less than one to nearly 200 cells/ml; the highest numbers occurred during the summer stratification of the water column. The ciliates contributed from 10 to 60 percent of the entire zooplankton biomass, suggesting that they have a substantial impact on the ecology of the lake. Beaver & Crisman (1982) made quantitative and qualitative comparisons of the ciliate faunas of a number of lakes in Florida characterized by different degrees of eutrophication (quantified on the basis of chlorophyll content). Ciliate concentrations varied from less than 10 to around 200 cells/ml and there was a clear correlation between ciliate numbers and the degree of eutrophication. They also found qualitative differences among the lakes. In the most oligotrophic lakes, the ciliate fauna was dominated by oligotrich forms which feed on phytoplankton cells, whereas in the most eutrophic ones, bacterivorous scuticociliates played a considerable quantitative role.

In addition to flagellates and ciliates, sarcodines also occur in the epilimnion of lakes. The testacean amoeba *Difflugia* is often reported to occur in large numbers in plankton (e.g., Pace, 1982; Schönborn, 1962). Also naked amoebae, including a form which feeds on algae (see Canter, 1980), and heliozoa are found in such lakes.

During summer stratification of both monomictic and meromictic lakes there is considerable production of chemolithotrophic bacteria at the chemocline (energy being obtained from the oxidation of sulfide, ammonia, and methane), while immediately beneath the chemocline, photosynthetic purple and green sulfur bacteria are active. These bacteria are predominantly consumed by protozoa which show a pronounced maximum in numbers at the chemocline (Sorokin, 1965; see also Figure 9.3). The reason for their dominant role in this limnic zone is not only due to the fact that protozoa are the most efficient consumers of bacteria, but also to the fact that many protozoan species tolerate or even require

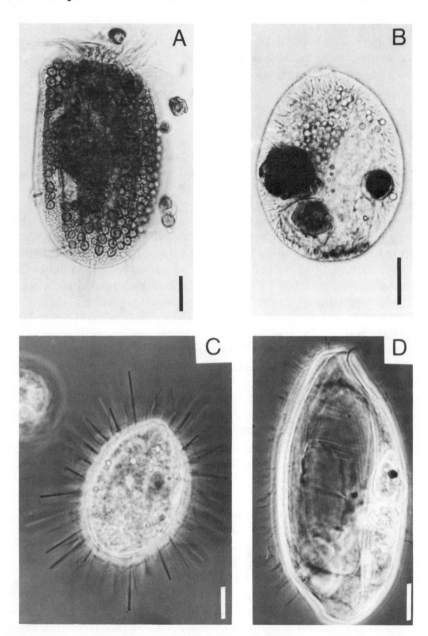

**Figure 9.2**  Ciliates from lake plankton. A. *Euplotes diadaleos* with zoochlo-rella; B. *Acaryophrya* sp. with ingested dinoflagellates; C. *Actinobolina vorax* with tentacles used for catching prey organisms; D. The predatory *Lembadion magnum*, the mouth of which fills nearly the entire ventral side. All scale bars: 20 μm. (All from the English Lake District and photographed by Hilda Canter-Lund.)

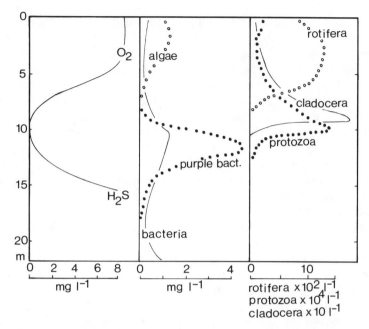

**Figure 9.3**   Vertical distribution of oxygen, sulfide, and some groups of organisms in Lake Belovod, USSR (redrawn from Sorokin, 1965).

anoxic or microaerophilic environments which metazoa cannot tolerate (Finlay, 1985).

The vertical zonation of chemical factors and microbial processes in stratified lakes closely resembles that which occurs in marine sediments (Section 8.3 and Figure 8.5); however, in the former case the zonation pattern spans meters rather than millimeters. Also, different protozoan species display zonation patterns which correlate with the patterns of chemical zonation, as is the case in marine sands (cf., Figure 8.8). Quite different protozoan communities are found in the lower part of the oxycline and in the anoxic hypolimnion than are found in the epilimnion (Bark, 1981). In all cases studied, however, the ciliate faunas at and below the oxycline mainly consist of benthic species which move up into the water column during summer stratification. This interesting fauna will be discussed in more detail in the following section.

## 9.3   Sediments, detritus, and solid surfaces

The bottom of ponds and lakes is often covered by algae, vascular plants, leaf litter, flocculent detrital matter, or patches of organisms, such as sulfur bacteria, cyanobacteria, or fungi. The surface sediment may there-fore provide a multitude of microhabitats for protozoa. These niches will be discussed together here because they harbor a large number of similar species; moreover, the extent to which microhabitat selection by different species occurs is not well documented, mainly due to the lack of sophistication of most sampling methods. It is in such benthic habitats that the most well known and familiar protozoa occur—the *Paramecium* species, *Tetrahymena, Stentor, Blepharisma, Stylony-chia, Actinosphaerium*, and *Amoeba proteus*. Much is known about these organisms because they have been extensively studied in the lab-oratory. However, these forms are not necessarily very common or im-portant in nature; rather they have been selected for laboratory studies because a) they are very large; b) they often appear in large numbers in samples from nature which have been enriched with organic material; or c) because they are easy to grow axenically.

Direct examination of samples of algae, sediment, detrital material, or surface scum from freshwater often reveals a large number of species. These include heterotrophic flagellates (euglenids, bodonids, chryso-monads, and flagellated amoeboid forms, often referred to as *Cerco-bodo*), a variety of amoebae, among which *Amoeba proteus* (Figure 2.3,A) is generally rare while the large *Pelomyxa* (Figure 2.3,B,C) some-times occurs in great numbers in flocculent sediments, a variety of tes-tacea (Figure 9.4), and of heliozoa (Figure 2.3,D-F). Ciliates also show a high diversity: among forms feeding on larger food particles (espe-cially photosynthetic organisms such as dinoflagellates and diatoms) species of *Frontonia* and various prostomatids are common, while spe-cies of *Nassula* specialize on filamentous Cyanobacteria. In microaer-ophilic environments, including decaying leaves, and within layers of detritus, species of *Loxodes* fill the food niche. This latter type of en-vironment also harbors large heterotrich ciliates belonging to the genera *Spirostomum* (Figure 9.5,A) and *Blepharisma*; these are both filter feeders depending on large bacteria and small eukaryotic cells. Ubiq-uitous in limnic habitats are hypotrich ciliates (e.g., species of *Stylony-chia, Holosticha, Keronopsis, Euplotes*, and *Aspidisca*). The bacteri-vorous ciliates found in lakes include scuticociliates, colpodid ciliates, *Paramecium*, and tetrahymenine ciliates, such as *Colpidium, Glau-coma*, and *Tetrahymena*. Carnivorous ciliates (depending on other cil-iates or on rotifers for food) include *Didinium, Lacrymaria, Spathi-dium, Dileptus, Trachelius*, and *Litonotus*. The giant colpodid, *Bursaria*, is a filter feeder which depends on large-sized prey, especially other ciliates. Species of *Ophryoglena* and, to some extent, *Coleps* are common histophagous ciliates in freshwater.

Associated with solid surfaces such as leaves of plants and algal fila-ments are a variety of colonial or solitary peritrichs (Figure 9.5)., spe-

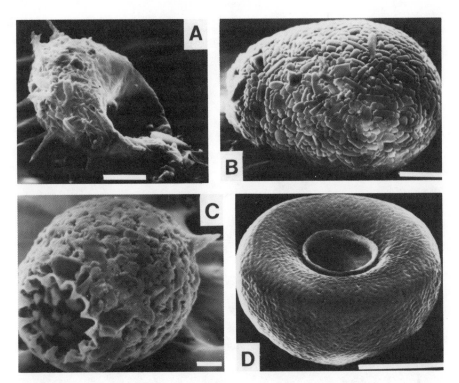

**Figure 9.4**   Scanning-electron micrographs of the tests of some freshwater testate amoebae. A. *Centropyxis spinosa*; B. *Nebela* sp.; C. *Difflugia corona*; D. *Arcella hemisphaerica*: all collected from ponds in Denmark. All scale bars: 20 μm.

cies of *Stentor*, and suctorians. Among the peritrichs, the most remarkable is undoubtedly *Ophrydium*: This zoochlorella-containing organism forms spherical and gelatinous colonies which may measure several centimeters across. They are attached to stones, plants, and other objects, and are common in many ponds and lakes, but their ciliate nature is probably often overlooked by limnologists (Winkler & Corliss, 1965; Foissner, 1979).

A true anaerobic fauna of ciliates also occurs in freshwater sediments where they are mainly represented by the same genera as those found in marine, anaerobic environments (e.g., *Metopus, Caenomorpha, Plagiopyla*, and representatives of the odontostomatids). In loose, flocculent or detrital sediments of lakes they are often found only a few millimeters beneath the surface and are consequently sampled together with aerobic forms.

The benthic ciliate fauna of lakes has been studied most thoroughly in the English Lake District and in a eutrophic lake in Scotland (Bark,

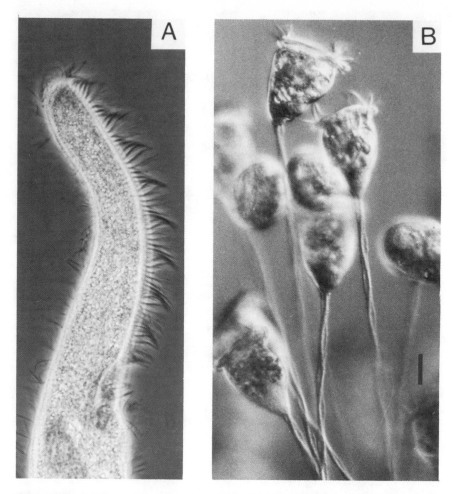

**Figure 9.5**  A. The anterior end of a *Spirostomum* cell showing the metachronal waves of the membranelle zone. B. *Vorticella* sp., scale bar: 20 µm. (A: from the English Lake District, photographed by Hilda Canter-Lund; B: from a Danish pond.)

1981; Finlay, 1978, 1980, 1981, 1982; Finlay et al., 1979; Goulder, 1974; Webb, 1961). In these lakes the density of ciliates is very high, on the order of several thousand cells/cm² and in aerobic sediments their numbers peak during the summer period. These numbers are in the upper range of that found in marine sediments. There is also a seasonal variation in the composition of species; while some species occur throughout the year others occur only or mainly during the summer. Altogether nearly a hundred species have been found in these lakes.

The chemical zonation of lake sediments resembles that described for marine sediments (Section 8.3). During the summer period, reducing conditions and the presence of sulfide prevail close to the sediment surface while during the winter period the sediments may be oxidized to a depth of several centimeters. Due to the flocculent nature of the surface layers of the sediment, protozoa may penetrate to a depth of at least six centimeters. However, in accordance with the oxidation-reduction properties of the sediment the cells concentrate within the upper centimeter during summer whereas the vertical zonation is much less compressed during the winter (Finlay, 1980).

In the eutrophic Esthwaite Water the sediments harbor a diverse fauna of ciliates from October (mixis) to May when the water column stratifies and the bottom layer of water becomes anoxic. During this period the ciliates number about 3500 cells/ml and the fauna is dominated by species belonging to *Loxodes* and *Spirostomum* (Figure 9.6). During the period of summer stratification some species disappear, while others enter the water column. This latter course is adopted by the two species of *Loxodes* which are found in the lower part of the oxycline at $O_2$ tensions of between zero and ten percent atmospheric saturation (Figure 9.7). The mechanisms by which *Loxodes* cells orient themselves in vertical oxygen gradients are discussed in Section 2.3. The statoliths of these geotactic ciliates (see Figure 1.3) contain barium. *Loxodes* is so abundant and there is so much barium in each cell, that most of the particulate barium in the lake water occurs inside these ciliates (Finlay, Hetherington & Davison, 1983). Some of the obligate anaerobes also leave the sediment and move to the anoxic hypolimnion.

Other accounts of the benthic protozoan communities of freshwater protozoa are by Fauré-Fremiet (1951) who studied patches of *Beggiatoa*, by Picken (1937) who studied protozoan communities associated with mats of cyanobacteria and fungi, and by Grolière & Njiné (1973) who studied ciliate populations in a forest pond over the course of one year. Fenchel (1975) studied protozoa in the detritus layer on the bottom of a pond on the arctic tundra. This study includes a quantification of heterotrophic microflagellates which numbered around $2 \times 10^5$ cells/ml of sediment. The author estimated that these protozoa consumed about 10 percent of the standing stock of bacteria per day, whereas the ciliates played only a minor role as bacterial consumers.

## 9.4 Running waters

Streams and rivers cannot sustain plankton populations because the cells are subject to "wash out," that is, they are carried downstream at a much higher rate than can possibly be compensated for by growth. Most

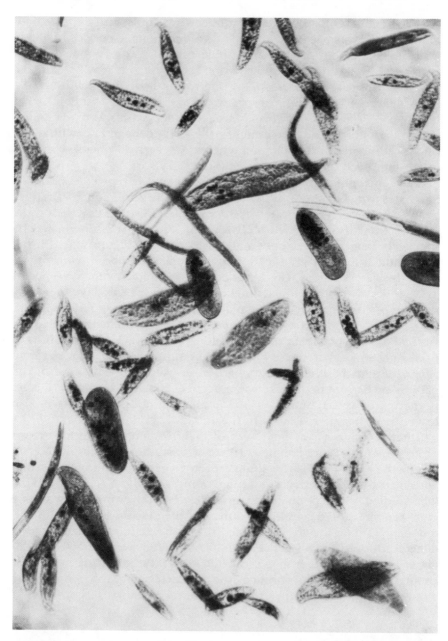

**Figure 9.6** A concentrated sample of the ciliate fauna of the benthos of Esthwaite Water, English Lake District, before summer stratification. The fauna is dominated by the about-200-μm-long *Loxodes striatus*; also seen are *L. magnus*, and *Frontonia* and *Spirostomum* spp. (Photographed by Bland J. Finlay.)

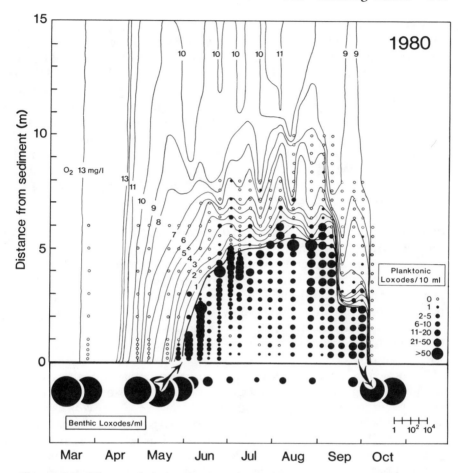

**Figure 9.7**  The vertical distribution of oxygen and of the ciliate *Loxodes* in Esthwaite Water, English Lake District, over a one-year period. (From Finlay & Fenchel, 1986.)

suspended cells found in the water of primary streams must originally have been dislodged from solid surfaces or sediments. A reservoir (a lake or pond) which receives water from the stream will be able to sustain pelagic populations, if the water is sufficiently mixed and if the "dilution rate" (the rate of outflow divided by the total water volume of the reservoir) does not exceed the growth rate of the planktonic organisms. The outflow from such a reservoir will, of course, contain plankton cells which may be able to grow in the stream or river water on their way to the next lake or, eventually, to the sea.

In running waters, protozoan populations may become established in sediments (typically, coarse gravel in rapid streams and finer sediments in rivers) and on solid surfaces such as rocks and aquatic plants. On such solid surfaces water flow becomes an important ecological factor because cells will be exposed to drag forces and thus to the risk of being dislodged. The physics of this problem is discussed by Silvester & Sleigh (1985). At the surface of an object (e.g., a leaf or a rock) immersed in the moving fluid, the velocity will be zero. Moving away from the surface the fluid velocity will gradually increase until it reaches the bulk velocity. Within the velocity gradient viscous forces are important and this zone is referred to as the "boundary layer," which is arbitrarily defined as the region around the object where the flow velocity is less than some fraction of the bulk velocity (usually 90 or 99 percent). The thickness of the boundary layer depends on the bulk velocity (it decreases with increasing velocity) and on the distance from the leading edge of the object (it increases with distance). For realistic values of flow velocities in streams, the boundary layer will always exceed 1 to 2 mm and thus completely contain any protozoa attached to surfaces. The protozoa will therefore only be exposed to a flow velocity which is a fraction of the bulk flow velocity. They will experience a force parallel to the direction of the flow and also a torque since they are exposed to a shear flow. Obviously, an organism minimizes its risk of being dislodged by being very flat and by having a large area of contact with its substrate.

Baldock et al. (1983) studied the protozoan fauna of chalk streams in southern England. They found that protozoa were associated with the bottom sediments and with the dominant aquatic plant, *Ranunculus penicillatus*, and were found predominately on the plants' older leaves and in "leaf nodes" which included a small volume of static water. The study clearly demonstrated that protozoan populations (permanently attached peritrichs as well as motile protozoa associated with surfaces) mainly occur in sheltered microhabitats (such as the leaf nodes) and also that populations are largest at locations of lowest current velocity. Amoebae, flagellates, and small ciliates were found, primarily in the form of flat species (e.g., the amoebae *Vanella* and the ciliate *Chilodonella*). These findings all suggest the importance of the water current as an agent for dislodging cells. Baldock (1980, quoted from Silvester & Sleigh, 1985) found a negative correlation between the occurrence of *Vorticella* and stream velocity and that a river spate with current velocities of up to 70 cm/sec removed most of these peritrichs.

W.D. Taylor (1983) studied the colonization of peritrich ciliates on artificial substrates submerged in streams. He found that in streams with a stable flow regime the genera *Vorticella* and *Carchesium* predominated; both these forms have contractile stalks. In streams with periodically high current velocities, forms with noncontractile stalks such as *Epistylis* predominated. It would seem that development of thicker and more rigid stalks (but with the sacrifice of the ability to contract) is an adaptation to rapid water currents. This interpretation is supported by examination of the stalk structure of some epizoic peritrichs which are exposed to high current velocities (Chapter 11).

## 9.5   Organic enrichment and polluted waters

Samples from freshwater habitats to which some organic material (e.g., cereals or other plant components, peptone, and skimmed milk) has been added show characteristic successions of microorganisms over the following days and weeks. Such experiments, if carried out systematically, disclose the nature of food chains and also reflect what happens in nature in a patch of decomposable material, such as the remains of dead animals or plants, or in an area of sewage outflow. Some aspects of such successions are discussed in conjunction with the microbial communities of detrital plant material in Section 8.3, but certain other properties are only found when rapidly decomposable material is studied. The succession of ciliates protozoa has been particularly well studied in freshwater environments.

A substantial amount of experimental work has been carried out by Legner (1973, 1975) and especially by Bick (1964, 1973, and references therein) using different sources of freshwater and different types and amounts of added organic material. The general trend observed in these experiments is that an initial bloom of bacteria is followed by the appearance of masses of heterotrophic flagellates and shortly thereafter by bacterivorous ciliates (e.g., *Colpidium, Glaucoma, Cyclidium*). The flagellates and the ciliates together reduce the number of bacteria substantially. Within a few days other ciliates which feed on larger particles, such as hypotrich ciliates and carnivorous pleurostomatids and prostomatids, appear. These are eventually followed by rotifers, small crustaceans, and other small metazoa. In general, the successions show a trend toward larger prey and larger predators. Eventually the organic material is totally decomposed and all populations decline. Under some circumstances, the systems tend to show predator-prey population oscillations. If the organic material contains nitrogen and phosphorus and if the material is exposed to light, populations of cyanobacteria and diatoms or other algae will appear which again support other protozoan populations. If the initial addition of organic material is sufficiently large, the systems may become anoxic and populations of anaerobic protozoa will develop.

Interest in this type of experiment has in part been sustained by attempts to use protozoa as indicators of aquatic pollution, in particular with respect to organic material. Around a point source of an organic pollutant the temporal successional patterns described above will appear as a spatial zonation pattern when moving away from the source of the pollution. The idea of using protozoan faunas as indicators of pollution goes back to the beginning of the century (Kolkwitz & Marsson, 1909) and the use of protozoa in "saprobiological indices" has been the subject of a vast amount of literature (e.g., Bick & Kunze, 1971, for ciliates, and Hänel, 1979, for heterotrophic flagellates). There is no doubt that massive additions of organic material to natural aquatic environments lead to dense populations of bacterivorous protozoa. There is also no doubt that the species composition of these communities reflects the chemical enviroment, especially if oxygen disappears and if hydrogen sulfide is present. Nevertheless, it would seem that the usefulness of protozoan communities for monitoring pollution with organic materials is overrated. The method is not more sensitive than simple chemical methods for quantifying such factors as dissolved or particulate organic carbon or ammonia and, especially in the case of hydrogen sulfide, the acute, human olfactory sense is more reliable and certainly simpler than the tedious business of identifying protozoa.

Activated sludge and percolating filters of sewage treatment plants develop special communities of protozoa which are sustained by the copious production of bacteria. Sarcodines, heterotrophic flagellates, and ciliates, as well as various small invertebrates occur, but ciliates have drawn the most attention. Population densities of protozoa are high; Curds (1973) mentions $5 \times 10^4$ cells/ml as typical of activated sludge. The diversity of species is also high; according to Curds (op.cit.) 53 species of ciliates were found in a percolating filter and 67 species were found in samples of activated sludge. The most abundant ciliates are the bacterivorous forms such as various tetrahymenine ciliates, scuticociliates, and attached peritrichs. Hypotrichs and predatory ciliates also occur. Faunistic accounts of the ciliates of sewage treatment plants are found in Curds & Cockburn (1970) and Madoni & Ghetti (1981).

Considerable attention has been given to the effect of the protozoan fauna on the overall efficiency of sewage treatment plants; this aspect has especially been studied by Curds (1973) and Curds et al. (1968, and references therein). These authors built small-scale, activated sludge plants in which protozoa were either present or absent. The results were quite clear: the presence of protozoa substantially reduced the amount of organic material, the viable count of bacteria, and the turbidity of the effluent. These results indicated that protozoa represent a necessary component in the efficient biological treatment of sewage. By considering the activated sludge process as a chemostat the authors illustrated the effect of protozoan grazing on bacteria and of protozoan predation on bacterivorous protozoa using both experimental and theoretical models. The basic result, that additional levels in the food chain (the protozoa in activated sludge probably contribute two to three) substantially enhance the efficiency of mineralization, can also be understood from the considerations given in Section 7.2.

# 10

# Protozoan Communities: Terrestrial Habitats

---

## 10.1 The nature of terrestrial protozoa

Protozoan motility, feeding, and growth require water; terrestral protozoa are therefore in a sense freshwater organisms living in the water which may either cover the surfaces of plants and litter as a thin film, or fill the pores of soils. In permanently flooded and saturated soils the protozoan fauna include species which are otherwise found in ponds, lakes, and streams. Well-drained terrestrial soils and litters, however, harbor protozoa with special adaptations for life in such environments and they rarely occur elsewhere (Stout, 1984). In this sense genuine terrestrial protozoa do exist.

The diversity of terrestrial protozoa has only recently been recognized and this is mainly due to previous methodological shortcomings. Soil protozoa are, with the exception of testaceans, very difficult to quantify or even observe directly in soil samples. Most studies have therefore relied on culture methods in which diluted soil samples are inoculated with bacterial suspensions. Such methods reveal only a small fraction of the species which may actually be present either as cysts or in an active state. This has been documented by Foissner (1982, 1985a,b, 1986) in particular: On the basis of direct microscopic observations of fresh soil samples, he found and described a large number of new, exclusively terrestrial ciliate species. However, a large number of new species of testate soil amoebae have also been found in recent years (e.g., Bonnet & Thomas, 1960, and references in Foissner, 1986).

Protozoa occur in soils, in litter layers, in mosses (the testacean fauna associated with *Sphagnum* has been especially well studied), and on the surfaces of leaves and other parts of terrestrial plants. Groups represented are naked and testate amoebae, heterotrophic flagellates, and ciliates. The terrestrial protozoa show adaptive traits, several of which are common to representatives of all these forms.

There are three environmental factors which are particularly important to terrestrial protozoa: the dimensions of pores in soils, the scarcity of water, and the temporal fluctuations in moisture. A nearly universal adaptive trait of terrestrial protozoa is their ability to produce resting cysts which remain viable in a desiccated state and which are formed in response to the disappearance of water. These protozoa generally excyst rapidly when water reappears. However, Volz (1972), who studied excystation in response to wetting, found that testacean cysts, when exposed to a level of moisture necessary for activity, excysted only after a period of several days. The period of time needed to excyst increased in proportion to the length of the previous period of dryness. It has also been found that some soil protozoa may encyst spontaneously at intervals (see Foissner, 1986). Both these features may be considered adaptive. Delayed excystation protects the cells from emerging during very brief periods of moisture (of shorter duration than one generation) and spontaneous encystation increases the probability of survival of clones exposed to very rapid and unpredictable episodes of desiccation of the habitat. It is possible that individual strains of soil protozoa show polymorphism with respect to such behavior (somewhat analogously to the delayed germination of seeds of annual plants in arid environments); however, no studies of this aspect of soil protozoa exist. The occurrence of reproductive cysts in some soil protozoa (e.g., the ciliate *Colpoda*) is also an adaptation to their temporally variable environment; following rainfall, bacterial numbers increase rapidly and the bacteria are then grazed by protozoa. Even if moisture continues to be present, food resources decline rapidly due to protozoan grazing and it is therefore competitively advantageous to feed while food is present while postponing reproduction, which requires encystation and then multiple cell divisions. The adaptive aspects of cyst formation are discussed further in Section 5.1. The temporal patchiness of the soil habitat is illustrated in Figure 10.1 which shows the changes in bacteria and bacterivorous amoebae following rainfall.

The most important groups of ciliates represented in terrestrial environments include the Colpodida (an order of ciliates which consists predominantly of terrestrial forms and which is likely to have evolved and diversified in terrestrial habitats), the Hypotrichida, and the Prostomatida. Common to the terrestrial representatives of these groups is a tendency for small cell sizes and for a slender or even vermiform shape. These traits represent an adaptation to life within the narrow crevices and pores of soils, reminiscent of the shapes of ciliates found in the

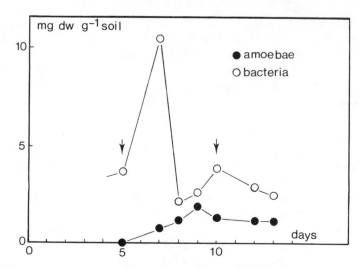

**Figure 10.1**   Biomass of bacteria and amoebae in a podsolized pine forest soil; arrows indicate times of rainfall. (Redrawn from Clarholm, 1981.)

interstitial spaces of marine sands (Section 8.3). The suspension-feeding ciliates of soils have a small mouth. This is a consequence of the viscous resistance to water flow in the vicinity of solid surfaces as was discussed in Section 3.1. Some characteristic, terrestrial ciliates are shown in Figure 10.2.

Testate amoebae show a number of morphological characteristics which have evolved independently within different taxonomic groups as an adaptation to terrestrial life (Bonnet, 1975; Foissner, 1986). As in ciliates, terrestrial testaceans tend to be smaller than their relatives living in ponds and lakes. Terrestrial forms have smooth tests, in contrast to freshwater forms which frequently carry spines or other ornamentations, and their tests are often flat or wedge-shaped; these traits are likely to facilitate movement within narrow crevices. The "ventral" side (where the opening for the pseudopodia is found) is often flattened while the opening itself is generally smaller than in freshwater forms. These traits may protect the cells against desiccation.

Naked amoebae of soils are usually small, but some large forms (whose pseudopodia may be several millimeters long) belonging to the poorly studied genus *Leptomyxa* also occur. Another large soil amoeba is *Archnula impatiens* which spreads its fine pseudopodial reticulum over an area of several mm². It feeds on protozoa, nematodes, and fungal spores. The family Vahlkampfiidae includes some important soil amoebae; among them species of *Vahlkampfia* are always amoeboid while the *Naegleria* species may be either amoeboid or flagellated (Section 5.1). Species of

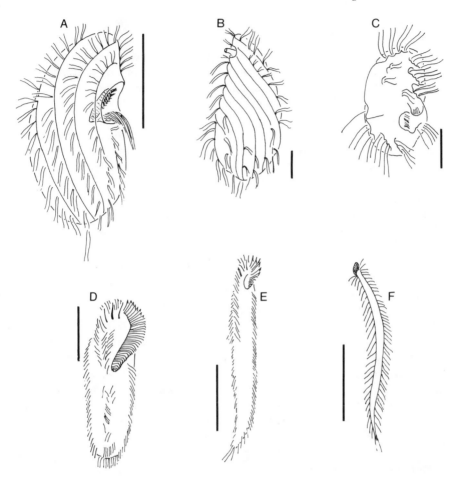

**Figure 10.2**   Examples of soil ciliates. A: *Colpoda steini*, B: *Grossglockneria acuta*, C: *Microthorax simulans*, D: *Holosticha muscorum*, E: *Amphisella acuta*, F: *Prospathidium bonetti*. Scale bars: A-C: 10 μm, D: 100 μm, E,F: 50 μm. (B: redrawn from Petz et al., 1985; C: redrawn from Foissner, 1985b; D,E: redrawn from Foissner, 1982; F: redrawn from Foissner, 1986.)

*Thecamoeba* are apparently unique among soil protozoa in not being capable of encystment, but the vegetative cells seem to survive desiccation (Old & Darbyshire, 1980; Page, 1976). The taxonomic composition of soil flagellates is very poorly documented; however, apparently nonpigmented chrysomonads and bodonids are common. The frequent occurrence of amoeboid flagellates of obscure taxonomic affinity which are often attributed to genera such as *Cercoboda* and *Mastigamoeba* seems to be a characteristic of soils.

Bacteria are undoubtedly the major food of the heterotrophic flagellates and the naked amoebae. The food niches of these two forms probably differ with respect to whether they feed on attached or on suspended bacteria and also on the basis of the size of the food particles. Testacean amoebae also include bacterivorous forms, but other species consume larger food such as protozoa and algal cells and some forms feed on fungal hyphae (Couteaux & Devaux, 1983). Among the ciliates, many of the smaller colpodids are filter feeders depending on suspended bacteria. The hypotrichs are also suspension feeders, but at least the larger forms feed on food items such as other protozoa, including testaceans. Most unique is the colpodid genus *Grossglockneria* and its relatives: Their mouths are developed into a tentacle which penetrates fungal hyphae and the ciliates feed by sucking the hyphal contents through the tentacle (Petz et al., 1985; Figure 10.2,B). Prostomatids, especially of the genus *Spathidium* and its relatives, constitute the most important ciliate predators in soils, feeding on other protozoa. The predatory ciliates belonging to the genus *Sorogena* which live in water films on the leaves of terrestrial plants were discussed in Section 5.1.

The minimum thickness of water films which can support active protozoa has attracted some attention. In soils, water is bounded by air menisci and soil particles. It is held by capillary forces and the strength of binding can be measured either as a "head" in cm $H_2O$ or as a pressure in dynes/cm$^2$. Soil biologists often express this force as the "water potential," pF, which is $\log_{10}$ (cmH$_2$O). The suction necessary to remove water from a soil sample is a function of capillary dimension; an approximate relationship between capillary diameter and pressure is: d (in $\mu$m) = 3/h, where h is measured in bars (1 bar = $10^6$ dynes/cm$^2$ = $10^3$ cm $H_2O$). If the amount of suction needed to remove water from a soil sample is plotted against its water content, the distribution of pore sizes in the soil can be estimated, since the largest pores empty first as the soil is dried (Russel, 1973).

From these considerations it is possible to determine the minimum size of water-filled pore which is necessary to sustain active protozoa. Darbyshire (1976) investigated this for the small ciliate *Colpoda steini* (Figure 10.2,A) and found that the ciliate is active only if the thickness of the water film is greater than about 30 $\mu$m. Using a somewhat different approach, Volz (1972) found that ciliates associated with litter are active when the water film exceeds about 50 $\mu$m. Small amoebae, however, can utilize smaller water-filled pores; based on experimental evidence Alabouvette et al. (1981) conclude that a water-film thickness of 3 $\mu$m constitutes the ultimate lower limit for protozoan activity. These results can be compared to the sizes of water-filled pores in ordinary soils at "field capacity," that is, with a water content which can be held against gravity; it is the state which can be obtained when a wetted soil is allowed to drain freely. An ordinary temperate soil at field capacity will contain 30 to 50 percent water and show a water suction of about 50 mbar, corresponding to pore sizes of about 60 $\mu$m; thus such a soil can sustain active protozoa. In tropical soils, however, the suction at field capacity may reach 350 mbar, corresponding to maximum sizes of water-filled pores of around 10 $\mu$m. If in such ordinary soils the water content is lowered to around 20 percent due to evaporation, only very fine pores are filled with water and protozoa become inactive. (The wilting point of plants is around 10 bar, corresponding to pore diameters of less than 0.3 $\mu$m and, depending on the type of soil, a water content of 10 to 20 percent.) In sandy soils, the water content at field capacity is much smaller (perhaps 3 to 8 percent since most of the pore space is composed of large pores which are emptied of water by gravity. The remaining fine, water-filled pores may be too scattered and isolated within the sand to allow protozoan activity.

There are several reports of the protozoan faunas of diverse types of terrestrial environments. Bamforth (1973) described the protozoan life on live and senescent leaves (see also Section 5.1) and Bamforth (1967) described the protozoan faunas of some subtropical forest soils. Foissner (1986) reviews the literature on protozoan soil faunas and demonstrates assemblages characteristic of different types of soils. The testacean faunas of carpets formed by the moss *Sphagnum* were studied in detail by Heal (1962, 1964a) and by Meisterfeld (1977). Both demonstrated vertical zonation patterns of various species which could be explained in part by different degrees of resistance to desiccation and in part by the fact that some testaceans harbor zoochlorella and therefore require light.

Faunas of soil protozoa are characterized not only by temporal but also by great spatial heterogeneity. This is due to the local occurrence of small patches with rapidly decomposing organic material, such as fecal material which harbors unique protozoan assemblages. These "coprophilic protozoa" have received some attention in the literature (e.g., Watson, 1946) and the topic has developed into a rather specialized "subdiscipline" (e.g., Smith, 1973).

There is a large amount of literature dealing with the changes (both quantitative and qualitative) in protozoan faunas which follow treatment of soils with fertilizers, pesticides, herbicides, or mineral oil, and artificial compression of soils. This literature is of limited interest in the present context and some of it suffers from methodological shortcomings. Most of it is critically reviewed by Foissner (1986).

## 10.2   The role of protozoa in soil ecosystems

The impact of protozoa on soil microbes and thus indirectly on soil fertility was a matter of debate at the beginning of the twentieth century and inspired a long series of experiments designed to elucidate the relationship between soil protozoa and bacteria which were performed at the Rothamsted Experimental Station in England (Cutler & Crump, 1920; Cutler et al., 1922; Cutler, 1927). These studies showed that protozoa (in particular the small flagellates and amoebae) control bacterial populations in soils and they demonstrated the rapid fluctuations in population sizes characteristic of both bacteria and protozoa. Since then, several other studies have been carried out in order to estimate the productivity of soil protozoa and their effect on the turnover of mineral nutrients. Much of this literature is reviewed by Stout (1980).

Such studies depend on a reliable method of quantifying protozoa in natural soil samples. To this end, dilution culture methods for estimating "most probably numbers" have been used almost exclusively although such methods have severe limitations. First of all, it is unlikely that all protozoa within a given soil sample will actually grow in a given culture medium; the work of Foissner (1986) suggests that the majority of the ciliate species do not multiply under the culture conditions usually employed. In studies of the productivity and metabolic activity of protozoa it is of interest to count only the active protozoa and not to count cysts. To obtain such figures, it is customary to treat a part of the soil sample with dilute hydrochloric acid, which supposedly kills active organisms but not cysts, and the number of active cells is then determined as the difference between the estimates of cell numbers for the untreated and the acid-treated samples. However, there is evidence that some cysts are in fact rendered nonviable by the acid treatment; if this is so and if the number of cysts is large relative to the number of active cells in the soil, the estimate of the latter may become very inflated relative to the true number (Foissner, 1986). The various methodologies used for quantifying soil protozoa are described and compared in Alabouvette et al. (1981) and in Stout et al. (1982).

When evaluating numbers of protozoa reported for different types of soils, it must also be kept in mind that the populations fluctuate considerably in size at different time scales, due to changes in factors such as moisture content, prey-predator population interactions (e.g., Figure 10.1), and seasonal changes in environmental conditions. Samples taken only on single occasions may therefore be of limited value if used to compare different types of soils or to estimate the average activities of the populations.

Stout et al. (1982) reviewed estimates of numbers of naked amoebae, flagellates, ciliates, and testaceans in different types of soils. Naked amoebae are reported to occur in numbers ranging from $10^3$ to $10^7$ cells per g (dry weight) of soil; in some cases a single soil locality spans the entire range. Clarholm (1981) found up to $2 \times 10^6$ amoebae per g of soil (the peak value in Figure 10.1); she considers the amoebae to be the single most important group of soil protozoa, comprising about 95 percent of the total protozoan biomass (Clarholm, 1985a). This seems reasonable as the amoebae are the protozoan group which is best adapted to life in thin water films. In addition, Stout et al. (1982) quote studies which show that flagellates are found in numbers ranging from less than $10^3$ to $3 \times 10^6$ cells per g of soil, while the corresponding values for ciliates range from 10 to $10^5$ cells per g of soil. Testaceans occur in numbers ranging from $10^3$ to $2 \times 10^5$ cells per g of soil (based on direct counts). Foissner (1986), generalizing from many studies (which largely overlap with those considered by Stout et al., 1982) states that in soils, protozoan biomass is equivalent to about half the metazoan biomass (mainly nematodes, annelids, and arthropods), but due to the higher weight-specific metabolic rate of smaller organisms, the protozoa may have a two—to three-fold larger share in the carbon cycle. Numbers of protozoa clearly correlate with the productivity of the site; thus forest soils harbor the largest number (and diversity) of protozoa and agricultural lands harbor more protozoa than grasslands. Numbers are much higher in the rhizosphere than in the bulk soil (Clarholm, 1981, 1985a; Stout, 1980).

There are several estimates of the annual productivity of testacean amoebae. These are based on total counts of live cells and empty tests at regular time intervals and on estimates of the rate at which empty shells disappear (e.g., Lousier, 1985; Lousier & Parkinson, 1984; Schönborn, 1977, 1982). These studies suggest that a number of annual generations ranges from a minimum of 10 to 20 to a maximum of more than 50. While such estimates cannot be considered as very accurate, they indicate that testacean growth rates are low relative to other similarly sized protozoa (cf. Section 4.2) and that their contribution to the overall carbon cycling of soils is modest. These estimates of the generation time of natural populations of testaceans are supported by the culture experiments of Heal (1964b) who found generation times ranging from six to eleven days. The relatively slow growth of testaceans is also in accord with the fact that their populations do not show the short-term fluctuations characteristic of other protozoan populations; rather they are characterized by regular seasonal patterns with one or two population maxima per year.

Other types of protozoa almost certainly have a greater impact on the metabolic activity of soils than testaceans do, but there are no very convincing estimates of feeding or growth rates in natural populations. Stout (1980) suggested that the smaller protozoa have generation times of one to three days; this is quite consistent with data such as those shown in Figure 10.1, but flagellates and the smaller ciliates such as *Colpoda* have the potential to grow much faster than this and, as argued in Section 4.2, organisms with a given potential for growth in the laboratory will realize this potential in nature from time to time. Also data, such as those shown in Figure 10.1 or the Rothamsted data which Stout (op.cit.) used to estimate growth rates, do not give information on protozoan mortality (e.g., due to predation) and hence tend to overestimate generation time.

Attempts to study the impact of protozoa of soil ecosystems have also been made using various types of experimental model ecosystems, such as flower pots containing sterilized soils to which various microbial components can be added. Habte & Alexander (1978) studied the bacterial density necessary to maintain populations of the ciliate *Tetrahymena*. They found that the protozoan populations ceased to grow when bacterial densities had decreased to around $10^6$ cells per g of soil; this approach is interesting, but since *Tetrahymena* does not naturally occur in soil the results are of limited value.

Coleman et al. (1978, and references therein) conducted a large number of experiments with soils inoculated with only one species of bacterium with or without the presence of both an amoeba and a nematode which consume bacteria as well as protozoa. The experiments demonstrate how the presence of protozoa enhances remineralization of nitrogen and phosphorus by consumption and hence stimulation of metabolic activity and carbon mineralization by bacteria. Clarholm (1985a,b) studied similar experimental soil systems with bacteria, with or without protozoa and with or without living wheat plants. She found that net mineralization to produce inorganic nitrogen from soil organic matter was strongly enhanced in the presence of protozoa and living roots (or the addition of soluble sugars). Apparently the excretion of low molecular weight organic compounds by the root tips stimulates bacterial growth. The bacteria in turn increase the rate of degradation of nitrogenous soil organic matter and assimilate the nitrogen into bacterial biomass. Eventually, protozoan populations increase in response to the increase in bacterial numbers and as the bacteria are consumed their cell nitrogen is remineralized by the protozoa and taken up by the plant roots.

# 11

# Symbiotic Protozoa

It is probable that almost all metazoans host a variety of symbiotic protozoa. These symbionts are found on body surfaces (including ciliated epithelia) such as the gills of aquatic organisms and the exoskeleton of crustacea, and are also found in the gut lumen and both intra- and intercellularly in various tissues and organs. A comprehensive treatment of symbiotic protozoa (which include causative agents of major medical and veterinary concern such as malarial, trypanosomal, and amoebal diseases) is beyond the scope of this book. However, symbionts are probably descended from free-living forms which subsequently adapted to life in the special habitats constituted by animals, and such evolutionary events are still taking place. Accordingly there is not a sharp delimitation between free-living and symbiotic protozoa. A few species belonging to the genera *Naegleria* and *Acanthamoeba* constitute a well-known example: These organisms are free-living amoebae in freshwater habitats; on rare occasions, individual cells invade humans to cause extremely virulent and lethal infections (e.g., Page, 1976). The whole spectrum from free-living organisms to facultative or obligate (and lethal) parasites of freshwater invertebrates is represented among species of the ciliate genus *Tetrahymena* (Corliss, 1973; Batson, 1985). Other examples are sessile ciliates which may regularly attach themselves to aquatic invertebrates.

The purpose of this section is twofold. One is to discuss some adaptations among free-living protozoa which render them "preadapted" to a symbiotic life. (Preadapted is not meant in a teleological sense, but as an indication that certain adaptive traits fortuitously make some organisms likely to invade and specialize in quite novel and different

161

habitats.) The other point is that while symbionts in a sense differ only from free-living forms in that they use parts of animals as their habitats (although this may lead to very specialized adaptations), their habitats are very well defined. From the viewpoint of evolutionary ecology this is advantageous for the elucidation of general problems of the evolution of niches, speciation, and species richness. This is discussed with reference to the ciliate epifauna of amphipod crustaceans.

The origin of the most specialized parasitic protozoa (namely the "Sporozoa" which include four presumably unrelated groups of organisms) is not known and there is little basis for discussing the characteristics of their free-living ancestors or the traits which made them develop into obligatory symbionts. In many other cases, however, it is possible to understand the evolution of symbiotic protozoa by comparing them to their free-living relatives.

Herbivorous animals in most cases depend on microbial fermentation in some part of the gut for the utilization of structural carbohydrates which make up the bulk of their food. The ruminant mammals constitute the best-known example, but many other mammal groups (e.g., perissodactyls, dugongs, and some marsupials) have independently developed a similar fauna, as have herbivorous reptiles, sea urchins, termites, and other creatures. In all these examples, specialized protozoa are present in the gut; in most cases they feed on the bacteria present. In some cases more complex food webs have developed; for example, in termites and cockroaches some flagellates; and in ruminants some ciliates ingest plant particles which are degraded within the protozoan cells, possibly by endosymbiotic bacteria. In the following discussion we will consider the question: How did these faunas evolve?

The ancestors of intestinal protozoa were probably free-living forms which were accidentally ingested by animals and which managed to survive and grow in the intestinal environment. One prerequisite for this process must be the ability to survive and grow in anoxic environments, and so one could expect that intestinal protozoa derive from the relatively few free-living anaerobic protozoa which occur in aquatic sediments and sulfureta. This is well documented in the gut fauna of sea urchins, which includes species of the ciliate genus *Metopus* and representatives of the trichostome family Plagiopylidae. These ciliates are normally free living in sulfide-containing and anoxic sediments. Other sea urchin ciliates also belong to the Trichostomatida, but they are not closely related to any known free-living forms (for references, see Fenchel et al., 1977). The trichostomatids, which include anaerobic, free-living forms, have also given rise to intestinal protozoa of terrestrial animals: the various types of ciliates in the rumen and cecum of herbivorous animals and the genus *Balantidium* which is represented by many species in the guts of a variety of mammals, reptiles, and amphibia.

The diplomonad flagellates, which are wide-spread intestinal symbionts of a great variety of animals, also have free-living representatives characteristic of anaerobic environments. The origins of the Parabasalidea, the flagellate group which includes the symbiotic trichomonads and the hypermastigid flagellates of termites, is unknown, but may well be polyphyletic.

There are other specialized traits in free-living protozoa which render them preadapted to a symbiotic life. Histophagy in ciliates (the ability to invade the tissue of animals through wounds; see Section 3.2) has, at least in some groups (some tetrahymenids and scuticociliates), evolved from bacterivory in carrion. From histophagy it is a small step to circumvent the necessity of a wound as the entry point and to invade healthy individuals directly (Fenchel, 1968). Another histophagous group, the ophryoglenids, have given rise to the genus *Ichthyophthirius*, the members of which are parasites on the skin of fish (Canella & Rocchi-Canella, 1976). One characteristic of many parasites—the polymorphic life cycle in which special stages are devoted to feeding, reproduction, and dispersal— had already evolved in all the cases discussed in the free-living ancestors as an adaptation to patchy environments (Section 5.1). All of these, and other examples not discussed here, show that the origin of complex adaptations to life as a parasite can often be understood as arising from adaptations in free-living ancestors.

Many permanently or periodically sedentary protozoa attach to other aquatic organisms. This may be fortuitous and the choanoflagellate shown in Figure 9.1 may represent such a case; since many choanoflagellates and other sedentary protozoa tend to attach to solid surfaces and it may not matter whether they attach to suspended detritus or to algal cells. F.J.R. Taylor (1982), however, cites a number of examples of peritrich ciliates, choanoflagellates, and even of a tintinnid ciliate associated with various kinds of planktonic algae in which the associations seem to be obligatory. With regard to the body surfaces of aquatic animals, a vast number of such associations have been described, in particular involving ciliates, and there are probably many more unknown ones. These associations involve specialized groups which always occur on a narrower or wider range of hosts (from a taxonomic viewpoint) (e.g., the stalkless peritrichs belonging to the genera *Trichodina* and *Urceolaria* which live on ciliated epithelia of a variety of organisms, or the thigmotrich ciliates which mainly occur on the gills of molluscs) as well as forms with free-living congeners, such as colonial and solitary peritrichs, which may be more or less specialized to a particular habitat. Colonial peritrichs are very common on planktonic copepods, so much so that they could have a significant ecological impact, but this has never been studied.

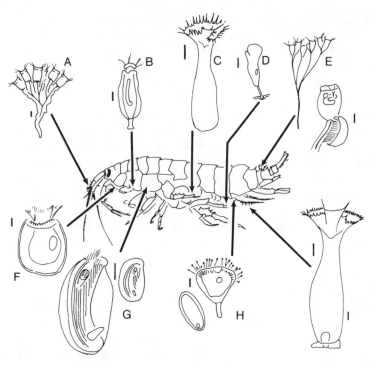

**Figure 11.1**    The ciliate epifauna on the amphipod *Gammarus locusta* in inner Danish Waters. A: *Zoothamnium* sp., B: *Cothurnia gammari*, C: *Heliochona sessilis*, D: *Conidophrys pilisuctor*, E: *Zoothamnium hiketes* with its ciliate parasite *Hypocoma parasitica*, F: *Lagenophrys* sp., G: two species belonging to the genus *Trochiloides*, H: *Heliochona scheuteni*, I: *Acineta foetida* and cysts of *Gymnodinioides inkystans*. All scale bars are 10 μm; the amphipod measures about 2 cm. (Redrawn from Fenchel, 1965.)

Among many examples only the ciliate faunas on amphipod crustaceans will be discussed here; these have been studied in detail (e.g., Fenchel, 1965; Nenninger, 1948; Precht, 1935, and a considerable amount of literature which treats the individual groups of ciliates; see Corliss, 1979) and they exemplify many characteristics of epizoic communities. Among these are a remarkable richness of species, very specialized habitat niches on the host, and a surprising host specificity. Figure 11.1 shows the ciliates which occur commonly in inner Danish waters, on the amphipod crustacean *Gammarus locusta*. These include four species of peritrichs (Figure 11.1,A,B,E, and F) which occur exclusively and respectively, on the antenna, the inner side of the side plates of the thorax, among the spines on the last abdominal segments, and on the maxillipeds. Two species of chonotrichs (a ciliate order only found as epizoans on crustaceans) are also found, one associated with the gills and the other on the pleopods (Figure 11.1,C,I). Both the chonotrichs and the peritrichs are filter feeders. A ciliate parasite of peritrichs, *Hypocoma*, is also found. In addition, the ciliate fauna includes two dysteriid ciliates (Figure 11.1,G) which feed on filamentous bacteria attached to the host surface, and a suctorian, *Acineta*, which presumably catches other ciliates (Figure 11.1,H). More intimately associated with the host (in the sense that they derive their nourishment from the host tissue) are two apostome ciliates, *Conidiophrys* (Figure 11.1,D) and cysts of *Gymnodinioides* (Figure 11.1,H) which excyst when the host molts and feed on host tissue in the exuvium.

In northern European waters the genus *Gammarus* actually includes five sibling species which show some niche diversification with respect to salinity, substratum, and breeding periods, but two or three of these sibling species commonly coexist. Nevertheless, and although the gammarid species are difficult to identify, their ciliate faunas are not identical even within one sampling area, showing diversity in particular with respect to the colonial peritrichs (Fenchel, 1965). Adaptation to an epizoic life can also be seen a few hours prior to molting of the host when the ciliates form free-swimming swarmers (telotrochs in the peritrichs, ciliated buds in the case of the chonotrichs), after which they move to the new exoskeleton.

What are the reasons for this richness of species and high degree of habitat specialization? It should first be mentioned that amphipods (and other pericarids) have features which render them especially suited for epibiosis. They have a habit of "precopula" formation in which the males remain attached to females for several days prior to egg fertilization, which can take place only when the female molts. The fertilized eggs develop with a metamorphosis in a brood pouch. These host traits should facilitate the transfer of ciliates from one host individual to another. The free-swimming stages of these ciliates refuse to settle on anything other than a gammarid host when isolated in dishes containing seawater.

Not all amphipods harbor such an impressive ciliate fauna and this probably is due to the fact that the *Gammarus* species form extremely dense and widespread populations in shallow waters, whereas most other species of amphipods are rarer and form more isolated populations. This relationship between host population sizes and species richness of epizoic ciliates can also be observed within the genus *Gammarus*. The species *G. duebeni* mainly form smaller, isolated populations in marginal habitats, such as rock pools and around outflows of streams along shorelines, and it harbors significantly fewer species of ciliates than do the other gammarids. The difficulty of the symbionts transferring from one host individual to another and perhaps the constraint of the total "habitat volume" therefore set limits to the evolution and maintenance of a large numbers of species.

The well-defined habitats of the epizoic protozoa are easily observed and described. It is tempting to ask whether the description of habitat niches of free-living forms is, in fact, very crude and if they are much more specialized with respect to environmental heterogeneity than hitherto observed. This is not likely to be the case since observations both of nature and of pure cultures suggest that most free-living species are much more catholic with respect to their environmental requirements than symbiotic species are; thus it would seem that if free-living species of attached peritrichs, for example, would be as specialized regarding where they attach, there would exist many more species of them (although, of course, there is no way of telling just how many more would be expected).

The only feature of animal surfaces as a habitat which seems to explain the high rate of speciation upon them is the fact that they must represent a fairly stable and predictable environment. This is mainly so because of the behavior of the host itself which attempts to avoid extremes in environmental factors, such as salinity, temperature, and anoxia, and fatal events such as desiccation or the grazing of its surfaces by animals. If this is the case the epizoic protozoa constitute an example of the current belief that the evolution and maintenance of a biotic community of great diversity requires a physically stable, predictable, and nonseasonal environment (May, 1973; Pianka, 1978) as has been inferred as the reason for the species richness of tropical climates and of the deep sea.

# 12

# Concluding Remarks

In this book, I have tried to evaluate the natural history of protozoa in light of a unifying theme: the physical constraints of being small and unicellular in relation to the physical and biological environment. Although protozoa comprise a systematically diverse assemblage of organisms, they share a number of functional traits which make them more than just a group of "small animals."

In the context of evolutionary ecology, protozoa pose a number of questions which have not yet been answered. These include the distribution and adaptive significance of sexuality and the mechanism of speciation in protozoa. One particularly interesting question is whether large complexes of sibling species, such as documented for *Paramecium* and *Tetrahymena*, are of widespread occurrence among other protozoan taxa. Also, how did such complexes arise and how are they maintained in nature? Although many examples of complex life cycles have been discovered in protozoa there is still considerable scope for descriptive work and even more scope for comparative studies which will elucidate an understanding of the ecological conditions which led to their evolution.

Until quite recently, most ecology texts offered only a very limited treatment of the role of microorganisms in the function of ecosystems (with the exception of bacteria and fungi in soils and of unicellular algae in the plankton). This paralleled the classical evolutionary tree of biology textbooks with the unicellular organisms at the bottom of the tree and man and other mammals at the top. Also from an ecological viewpoint the protista (as well as the prokaryotes) were considered to be some sort of Precambrian relics which only marginally affect the properties and function of the extant biosphere. Nevertheless, it has since been established that prokaryotes play a qualitatively and quantitatively indispensable role in biogeochemical cycles and therefore also in the biosphere.

The ecological importance of protozoa is difficult to assess because their functional role in the carbon flow as phagotrophs is not qualitatively different from that of metazoa. However, the role of protozoa is perhaps best understood by considering their enormous entire size range. Figure 1.4 shows that all the protozoa together display a size range comparable to that constituted by the metazoa. This statement, of course, is based on the logarithmic scale of the figure, but this is reasonable in their ecological and functional context: Prey and predator species are characterized by a certain size ratio rather than a constant size difference and "short circuits" of food chains (that is, that very large organisms eating extremely small food particles) are very unusual. This property is explained by purely mechanical considerations as discussed in Chapter 3. Within any particular size group of organisms, different adaptations for catching food particles will allow for a certain range in food particle size, but as has been demonstrated, the range in relative prey sizes among ciliates (from about 1 to 0.01 times body size) is comparable to that found among mammals (comparing a tiger with a baleen whale). When the lengths of phagotrophic organisms are arranged on a logarithmic scale, this reflects the number of trophic levels of a biotic community (with approximately one level per decade). Protozoa therefore play an indispensable role in ecological systems in which the basis of the food chains consists of bacteria or minute primary producers.

It was shown in Chapter 7 that empirical evidence as well as theoretical reasoning suggest that in phagotrophic food chains, each logarithmic size class will roughly represent identical biomasses. Consequently, Figure 1.4 shows that in ecological systems in which the basis of the food chain is constituted by the most minute organisms, protozoa will be represented by a biomass comparable to that of metazoan animals and the former will be responsible for a considerably larger share of the carbon flow.

Another ecological aspect of size is the utilization of temporal and spatial patchiness of habitats and food resources. The scale of such environmental patchiness determines which size class of organisms will most efficiently exploit a patchy environment. It is perhaps trivial that this applies to spatial patchiness; it also applies to temporal patchiness since the rate constants of physiological processes tend to increase with decreasing body size. The heterogeneity of natural ecosystems thus creates ecological niches for all size classes of organisms.

In describing these characteristics of protozoa I hope that I have shown their ecological importance (in addition to their significance in cell biology, systematics, general natural history, and, of course, their esthetic appeal). At the same time I hope that in describing the natural history of protozoa I have contributed to our general understanding of the structure and function of biological communities.

# References

Agamaliev, F. G. 1967. Faune de ciliés mésopsammiques de al côte ouest de la Mer Caspienne. *Cah. Biol. Mar.* 8: 359–402.

Agamaliev, F. G. 1974. Ciliates of the solid surface overgrowth of the Caspian Sea (in Russian). *Acta Protozool.* 8: 53–83.

Alabouvette, C., Couteaux, M. M., Old, K. M., Pussard, M., Reisinger, O. & Toutain, F. 1981. Le protozoaires du sol: aspects écologiques et méthodologiques. *Ann. Biol.* 20: 255–303.

Allen, R. D. 1984. *Paramecium* phagosome membrane: from oral region to cytoproct and back again. *J. Protozool.* 31: 1–6.

Almagor, M., Ron, A. & Bar-Tana, J. 1981. Chemotaxis in *Tetrahymena thermophila. Cell Motility* 1: 261–268.

Amos, W. B. 1975. Contraction and calcium binding in vorticellid ciliates. Pp. 411–436 in R. E. Stephens & S. Inoue (eds.), *Molecules and Cell Movements.* Raven Press, New York.

Andersen, P. & Fenchel, T. 1985. Bacterivory by microheterotrophic flagellates in seawater samples. *Limnol. Oceanogr.* 30: 198–202.

Anderson, O. R. 1983. *Radiolaria.* Springer-Verlag, New York.

Anderson, O. R. & Bé, A. W. H. 1976a. A cytochemical fine structure study of phagotrophy in a planktonic foraminifer, *Hastegerina pelagica* (d'Orbigny). *Biol. Bull.* 151: 437–449.

Anderson, O. R. & Bé, A. W. H. 1976b. The ultrastructure of a planktonic foraminifer, *Globigerinoides sacculifer* (Brady), and its symbiotic dinoflagellates. *J. Foram. Res.* 6: 1–21.

Anderson, O. R., Spindler, M., Bé, A. W. H. & Hemleben, C. 1979. Trophic activity of planktonic foraminifera. *J. Mar. Biol. Assoc. U. K.* 59: 791–799.

Andresen, N., Chapman-Andresen, C. & Nilsson, J. R. 1968. The fine structure of *Pelomyxa palustris. C. R. Trav. Lab. Carlsberg* 36: 285–317.

171

Antipa, G. A., Martin, K. & Rintz, M. 1983. A note on the possible ecological significance of chemotaxis in certain ciliated protozoa. *J. Protozool.* 30: 55–57.

Ax, P. & Ax, R. 1960. Experimentelle Untersuchungen über die Salzgehaltstoleranz von Ciliaten aus dem Brachwasser und Süsswasser. *Biol. Zbl.* 79: 7–31.

Azam, F., Fenchel, T., Field, J. G., Gray, J. S., Meyer-Reil, L. A. & Thingstad, F. 1983. The ecological role of water-column microbes in the sea. *Mar. Ecol. Prog. Ser.* 10: 257–263.

Baldock, B. M., Baker, J. H. & Sleigh, M. A. 1983. Abundance and productivity of protozoa in chalk streams. *Holarctic Ecol.* 6: 238–246.

Bamforth, S. S. 1967. A microbial comparison of two forest soils of Southeastern Louisiana. *Proc. Louisiana Acad. Sci.* 30: 7–16.

Bamforth, S. S. 1973. Population dynamics of soil and vegetation protozoa. *Am. Zool.* 13: 171–176.

Bardele, C. F. 1972. A microtubule model for ingestion and transport in the suctorian tentacle. *Z. Zellforsch.* 126: 116–134.

Bardele, C. F. & Grell, K. G. 1967. Elektronenmikroskopische Beobachtungen zur Nahrungsaufname bei dem Suktor *Acineta tuberosa* Ehrenberg. *Z. Zellforsch.* 80: 108–123.

Bark, A. W. 1981. The temporal and spatial distribution of planktonic and benthic protozoan communities in a small productive lake. *Hydrobiologia* 85: 239–255.

Batson, B. S. 1985. A paradigm for the study of insect-ciliate relationships: *Tetrahymena sialidos* sp. nov. (Hymenostomatida: Tetrahymenidae), parasite of larval *Sialis lutaria* (Linn.) (Megaloptera: Sialidae). *Phil. Trans. R. Soc. Lond.* B 310: 123–144.

Bé, A. W. H., Hemleben, C., Anderson, O. R., Spindler, M., Hacunda, J. & Tuntivate-Choy, S. 1977. Laboratory and field observations of living planktonic foraminifera. *Micropalaeontology* 23: 155–179.

Beam, C. A. & Himes, M. 1980. Sexuality and meiosis in dinoflagellates. Pp. 171–206 in M. Levandowsky & S. H. Hutner (eds.), *Biochemistry and Physiology of Protozoa.* 2nd ed., vol. 3. Academic Press, New York.

Beauchop, T. & Elsden, S. R. 1960. The growth of microorganisms in relation to their energy supply. *J. Gen. Microbiol.* 23: 457–469.

Beaver, J. R. & Crisman, T. L. 1982. The trophic response of ciliated protozoans in freshwater lakes. *Limnol. Oceanogr.* 27: 246–253.

Berg, H. C. 1983. *Random Walks in Biology.* Princeton Univ. Press, Princeton.

Berger, J. D. & Pollock, C. 1981. Kinetics of food vacuole accumulation and loss in *Paramecium tetraurelia. Trans. Am. Microsc. Soc.* 100: 120–133.

Berk, S. G., Brownlee, D. C., Heinle, D. R., Kling, H. J. & Colwell, R. R. 1977. Ciliates as a food source for marine planktonic copepods. *Microb. Ecol.* 4: 27–40.

Bick, H. 1964. Die Sukcession der Organismen bei der Selbstreinigung von organisch verunreinigtem Wasser unter verschiedenen Milieubedingungen. *Min. ELF. des Landes Nordrhein/Westfalen.* Düsseldorf.

Bick, H. 1966. Populationsökologische Beobachtungen über das Auftreten sexueller Prozesse bei Süsswasserpolypen und Ciliaten. *Zool Anz.* 176: 183–192.

Bick, H. 1973. Population dynamics of protozoa associated with the decay of organic materials in freshwater. *Am. Zool.* 13: 149–160.

Bick, H. & Kunze, S. 1971. Eine Zusammenstellung von autökologischen und saprobiologischen Befunden an Süsswasserciliaten. *Int. Revue Ges. Hydrobiol.* 56: 337–384.

Biechler, B. 1952. Récherchés sur les Peridiniens. *Bull. Biol. France Belg.* (suppl.) 36: 1–149.

Blackbourn, D. J., Taylor, F. J. R. & Blackbourn, J. 1973. Foreign organelle retention by ciliates. *J. Protozool.* 20: 286–288.

Blakemore, R. P., Frankel, R. B. & Kalmijn, Ad. J. 1980. South-seeking magnetotactic bacteria in the Southern Hemisphere. *Nature, Lond.* 286: 384–385.

Bonnet, L. 1975. Types morphologiques, écologie et evolution de la théque chez les Thécamoebiens. *Protistologica* 11: 363–378.

Bonnet, L. 1979. Faunistique et biogéographie des thécamoebiens. V. Thécamoebiens de quelques sols du Brésil et du Paraguay. *Bull. Soc. Hist. Nat. Tolouse* 115: 119–122.

Bonnet, L. & Thomas, R. 1960. Thécamoebiens du sol. *Vie Milieu* (suppl.) 11: 1–103.

Borror, A. 1972. Tidal marsh ciliates (Protozoa): morphology, ecology, systematics. *Acta Protozool.* 10: 29–71.

Bradbury, P. C. & Olive, L. S. 1980. Fine structure of the feeding stage of a sorogenic ciliate, *Sorogena stoianovitchae* gen. n., sp. n. *J. Protozool.* 27: 267–277.

Brock, T. D. 1969. Microbial growth under extreme conditions. Pp. 15–41 in *Microbial Growth*. 19th Symposium for General Microbiology. Cambridge University Press, Cambridge.

Brock, T. D. 1978. *Thermophilic Microorganisms and Life at High Temperatures*. Springer-Verlag, New York.

Brown, J. A. & Nielsen, P. J. 1974. Transfer of photosynthetically produced carbohydrate from endosymbiotic chlorellae to *Paramecium bursaria. J. Protozool.* 21: 569–570.

Burkill, P. H. 1982. Ciliates and other microplankton components of a nearshore food-web: standing stocks and production processes. *Ann. Inst. Oceanogr., Paris* 58: 335–350.

Burkovsky, I. V. 1970. Ciliates of the sand littoral and sublittoral of Kandalaksha Gulf (White Sea) and the analysis on the fauna of benthic ciliates of other seas (in Russian). *Acta Protozool.* 8: 183–201.

Burnett, B. R. 1973. Observations of the microfauna of the deep-sea benthos using light and scanning electron microscopy. *Deep-Sea Res.* 20: 413–417.

Burnett, B. R. 1977. Quantitative sampling of microbiota of the deep-sea benthos—I. Sampling techniques and some data from the abyssal central North Pacific. *Deep-Sea Res.* 24: 781–789.

Burnett, B. R. 1981. Quantitative sampling of nanobiota (microbiota of the deep-sea benthos—III. The bathyal San Diego Trough. *Deep-Sea Res.* 28A: 649–663.

Burns, R. G. & Slater, J. H. (eds.) 1982. *Experimental Microbial Ecology*. Blackwell, Oxford.

Cachon, M. & Caran, B. 1979. A symbiotic green algae, *Pedinomonas symbiotica* sp. nov. (Prasinophyceae), in the radiolarian *Thalassolampe margarodes*. *Phycologia* 18: 177–184.

Cairns, J. Jr. & Ruthven, J. A. 1972. A test of the cosmopolitan distribution of fresh-water protozoans. *Hydrobiologia* 39: 405–427.

Calow, P. 1977. Conversion efficiencies in heterotrophic organisms. *Biol. Rev.* 52: 385–409.

Cameron, I. L. 1973. Growth characteristics of *Tetrahymena*. Pp. 199–266 in A. M. Elliott (ed.), *Biology of Tetrahymena*. Dowden, Hutchinson & Ross, Inc., Stroudsburg, Pennsylvania.

Canella, M. F. 1951. Contributions à la connaissance de gymnostomes des genres *Holophrya, Amphileptus* et *Litonotus* prédateurs de *Carchesium polypinum* et d'autres peritriches sessiles. *Ann. Univ. Ferrara N.S.* 1: 1–11.

Canella, M. F. 1957. Studi e ricerche sui tentaculiferi nel quadro della biologia generale. *Ann. Univ. Ferrara N.S.* 1: 259–716.

Canella, M. F. & Rocchi-Canella, L. 1976. Biologie des Ophryoglenina. *Ann. Univ. Ferrara N.S.* 3: 1–150.

Canter, H. M. 1980. Observations on the amoeboid protozoan *Asterocaelum* (Proteomyxida) which ingests algae. *Protistologica* 16: 475–483.

Cappuccinelli, P. 1980. *Motility of Living Cells*. Chapman and Hall Ltd, New York.

Capriuolo, G. M. & Carpenter, E. J. 1980. Grazing by 35 to 202 μm microzooplankton in Long Island Sound. *Mar. Biol.* 56: 319–326.

Capriuolo, G. M. & Ninivaggi, D. V. 1982. A comparison of the feeding activities of field collected tintinnids and copepods fed identical natural particle assemblages. *Ann. Inst. Oceanogr., Paris* 58: 325–334.

Caron, D. A., Davis, P. G., Madin, L. P. & Sieburth, J. McN. 1982. Heterotrophic bacteria and bacterivorous protozoa in oceanic macroaggregates. *Science* 218: 795–797.

Caron, D. A., Goldman, J. C., Andersen, O. K. & Dennett, M. R. 1985. Nutrient cycling in a microflagellate food chain: II. Population dynamics and carbon cycling. *Mar. Ecol. Prog. Ser.* 24: 243–254.

Caron, F. & Meyer, E. 1985. Does *Paramecium primaurelia* use a different genetic code in its macronucleus? *Nature, Lond.* 314: 185–188.

Chapman-Andresen, C. 1967. Studies on endocytosis in amoebae. The distribution of pinocytically ingested dyes in relation to food vacuoles in *Chaos chaos*. I. Light microscopic observations. *C. R. Trav. Lab. Carlsberg* 36: 161–187.

Chapman-Andresen, C. & Hamburger, K. 1981. Respiratory studies on the giant amoeba *Pelomyxa palustris*. *J. Protozool.* 28: 433–440.

Christiansen, B. 1958. The foraminifer fauna in Dröbak Sound in the Oslo Fjord (Norway). *Nytt Magasin Zool.* 6: 5–91.

Christiansen, B. O. 1964. *Spiculosiphon radiata*, a new foraminifera from Northern Norway. *Astarte* 25: 1–8.

Christiansen, B. O. 1971. Notes on the biology of foraminifera. *Vie Milieu. Suppl.* 22: 465.

Christopher, M. H. & Patterson, D. J. 1983. *Coleps hirtus*, a ciliate illustrating facultative symbioses between protozoa and algae. *Ann. Stat. Biol. Besse-en-Chandesse* 17: 278–297.

Clarholm, M. 1981. Protozoan grazing of bacteria in soil—impact and importance. *Microb. Ecol.* 7: 343–350.

Clarholm, M. 1985a. Possible roles for roots, bacteria, protozoa and fungi in supplying nitrogen to plants. Pp. 355–365 in A. F. Fitter (ed.), *Ecological Interactions in Soil*. Blackwell, Oxford.

Clarholm, M. 1985b. Interactions of bacteria, protozoa and plants leading to mineralization of soil nitrogen. *Soil. Biol. Biochem.* 17: 181–187.

Coleman, D. C., Cole, C. V., Hunt, H. W. & Klein, D. A. 1978. Trophic interactions in soils as they affect energy and nutrient dynamics. I. Introduction. *Microb. Ecol.* 4: 345–349.

Coleman, G. S. 1979. Rumen ciliate protozoa. Pp. 381–408 in M. Levandowsky & S. H. Hutner (eds.), *Biochemistry and Physiology of Protozoa*. 2nd ed., vol. 2. Academic Press, New York.

Corliss, J. O. 1972. Current status of the international collection of ciliate type specimens and guidelines for future contributors. *Trans. Amer. Microsc. Soc.* 91: 221–235.

Corliss, J. O. 1973. History, taxonomy, ecology, and evolution of species of *Tetrahymena*. Pp. 1–55 in A. M. Elliott (ed.), *Biology of Tetrahymena*. Dowden, Hutchinson & Ross, Inc., Stroudsburg, Pennsylvania.

Corliss, J. O. 1979. *The Ciliated Protozoa*. Pergamon Press, Oxford.

Corliss, J. O. 1984. The kingdom Protista and its 45 phyla. *BioSystems* 17: 87–126.

Corliss, J. O. & Daggett, P.-M. 1983. "*Paramecium aurelia*" and "*Tetrahymena pyriformis*": current status of the taxonomy and nomenclature of these popularly known and widely used ciliates. *Protistologica* 19: 307–322.

Corliss, J. O. & Esser, S. C. 1974. Comments on the role of the cyst in the life cycle and survival of free-living protozoa. *Trans. Amer. Microsc. Soc.* 93: 578–593.

Couteaux, M. M. & Devaux, J. 1983. Effet d'un enrichissement en champignon sur la dynamique d'un peuplement thécamoebien d'un humus. *Rev. Écol. Biol. Sol.* 20: 519–545.

Cronkite, D. & Van den Brink, S. 1981. The role of oxygen and light in guiding "photoaccumulation" in the *Paramecium bursaria-Chlorella* symbiosis. *J. Exp. Zool.* 217: 171–177.

Curds, C. R. 1973. The role of protozoa in the activated-sludge process. *Am. Zool.* 13: 161–169.

Curds, C. R. & Cockburn, A. 1970. Protozoa in biological treatment processes—I. A survey of the protozoan fauna of British percolating filters and activated-sludge plants. *Water Res.* 4: 225–236.

Curds, C. R., Cockburn, A. & Vandyke, J. M. 1968. An experimental study of the role of the ciliated protozoa in the activated-sludge process. *Wat. Pollut. Control.* 67: 312–329.

176    References

Cutler, D. W. 1927. Soil protozoa and bacteria in relation to their environment. *J. Queckett. Microsc. Club* 15: 309–330.

Cutler, D. W. & Crump, L. M. 1920. Daily periodicity in the numbers of active soil flagellates: with a brief note on the relation of trophic amoebae and bacterial numbers. *Ann. Appl. Biol.* 7: 11–24.

Cutler, D. W., Crump, L. M. & Sandon, H. 1922. A quantitative investigation of the bacterial and protozoan population of the soil, with an account of the protozoan fauna. *Phil. Trans. R. Soc. London,* Ser. B 211: 317–350.

Dando, P. R., Southward, A. S., Southward, E. C., Terwilliger, N. B. & Terwilliger, R. C. 1985. Sulphur-oxidizing bacteria and haemoglobin in gills of the bivalve mollusc *Myrtea spinifera. Mar. Ecol. Prog. Ser.* 23: 85–98.

Darbyshire, J. F. 1976. Effect of water suctions on the growth in soil of the ciliate *Colpoda steini* and the bacterium *Azotobacter chroococcum. J. Soil. Sci.* 27: 369–376.

Davidson, L. A. 1982. Ultrastructure, behavior and algal flagellate affinities of the helioflagellate *Ciliophrys marina,* and the classification of the helioflagellates (Protista, Actinopoda, Heliozoa). *J. Protozool.* 29: 19–29.

Davis, P. G., Caron, D. A. & Sieburth, J. McN. 1978. Oceanic amoebae from the North Atlantic: culture, distribution, and taxonomy. *Trans. Amer. Microsc. Soc.* 97: 73–88.

Davis, P. G. & Sieburth, J. McN. 1982. Differentiation of phototrophic and heterotrophic nanoplankton populations in marine waters by epifluorescence microscopy. *Ann. Inst. Océanogr., Paris* 58: 249–260.

Dawson, J. A. & Hewitt, D. C. 1931. The longevity of encysted colpodas. *Am. Nat.* 65: 181–186.

Deroux, G. 1976. Plan corticale des cyrtophorida III—Les structures differenciatrices chez les dysterina. *Protistologica* 12: 505–538.

Dietz, G. 1964. Beitrag zur Kenntnis der Ciliatenfauna einiger Brackwassertümpel (Etangs) der französischen Mittelmeerküste. *Vie Milieu* 15: 17–93.

Dingfelder, J. H. 1962. Die Ciliaten vorübergehender Gewässer. *Arch. Protistenk.* 105: 509–658.

Dini, F. 1981. Relationship between breeding systems and resistance to mercury in *Euplotes crassus* (Ciliophora: Hypotrichida). *Mar. Ecol. Prog. Ser.* 4: 195–202.

Dini, F. 1984. On the evolutionary significance of autogamy in the marine *Euplotes* (Ciliophora: Hypotrichida). *Am. Nat.* 123: 151–162.

Dini, F. & Luporini, P. 1982. The inheritance of the mate-killer trait in *Euplotes crassus* (Hypotrichida, Ciliophora). *Protistologica* 13: 179–184.

Dogiel, V. A. 1965. *General Protozoology.* At the Clarendon Press, Oxford.

Dragesco, J. 1960. Les ciliés mésopsammiques littoraux. *Trav. St. Biol. Roscoff (N.S.)* 12: 1–336.

Dragesco, J. 1962. Capture et ingestion des proies chez les infusoires ciliés. *Bull. Biol. France Belg.* 46: 123–167.

Dragesco, J. 1963. Revision du genre *Dileptus* Dujardin 1871 (Ciliata Holotricha) (systematique, cytologie, biologie). *Bull. Biol. France Belg.* 97: 103–145.

Dragesco, J. 1964. Capture et ingestion des proies chez *Actinosphaerium eichorni* (Rhizopoda, Heliozoa). *Arch. Zool. Exp. Gén.* 104: 163–175.

Dragesco, J. 1965. Étude cytologique de quelques flagellés mésopsammiques. *Cah. Biol. Mar.* 6: 83–115.

Dragesco, J. 1968. A propos de *Neobursaridium gigas* Balech 1941: Sténothermie, inclusions, ultrastructure des trichocystes. *Protistologica* 4: 85–106.

Dragesco, J. & Dragesco-Kerneis, A. 1986. *Ciliés libres de l'Afrique intertropicale*. Édition de l'Orstrom, Paris.

Droop, M. R. 1966. The role of algae in the nutrition of *Heteramoeba clara* Droop with notes on *Oxyrrhis marina* Dujardin and *Philodina roseola* Ehrenberg. Pp. 269–282 in H. Barnes (ed.), *Some Contemporary Studies in Marine Science*. Allen & Unwin, London.

Ducklow, H. W. 1983. Production and fate of bacteria in the oceans. *BioScience* 33: 494–501.

Ehrenberg, C. G. 1838. *Die Infusionsthierchen als vollkommene Organismen*. Leipzig.

Elliott, A. M. 1973. Life cycle and distribution of *Tetrahymena*. Pp. 259–286 in A. M. Elliott (ed.), *Biology of Tetrahymena*. Dowden, Hutchinson & Ross, Stroudsburg, Pennsylvania.

Erez, J. 1978. Vital effect on stable isotope composition seen in foraminifera and coral skeletons. *Nature, Lond.* 273: 194–202.

Fauré-Fremiet, E. 1948. Le rythme de marée du *Strombidium oculatum* Gruber. *Bull. Biol. France Belg.* 82: 3–23.

Fauré-Fremiet, E. 1950a. Caulobacteries epizoiques associées aux *Centrophorella* (Ciliés holotriches). *Bull. Soc. Zool. Fr.* 75: 134–137.

Fauré-Fremiet, E. 1950b. Écologie des ciliés psammophiles littoraux. *Bull. Biol. France Belg.* 84: 35–75.

Fauré-Fremiet, E. 1951. Associations infusoriennes à *Beggiatoa*. *Hydrobiologia* 3: 65–71.

Febvre, J. & Febvre-Chevalier, C. 1979. Ultrastructural study of zoxanthellae of three species of Acantharia (Protozoa: Actinopoda), with details of their taxonomic position in the Prymnesiales (Prymnesiophyceae, Hibberd, 1976). *J. Mar. Biol. Assoc., U.K.* 59: 215–226.

Febvre-Chevalier, C. & Febvre, J. 1982. Locomotion processes in some pelagic and benthic protozoa. *Ann. Inst. Océanogr., Paris* 58: 137–142.

Fenchel, T. 1965. On the ciliate fauna associated with the marine species of the amphipod *Gammarus* J. G. Fabricius. *Ophelia* 2: 281–303.

Fenchel, T. 1968. The ecology of marine microbenthos. II. The food of marine benthic ciliates. *Ophelia* 5: 73–121.

Fenchel, T. 1969. The ecology of marine microbenthos. IV. Structure and function of the benthic ecosystem. *Ophelia* 6: 1–182.

Fenchel, T. 1970. Studies on the decomposition of organic detritus derived from the turtle grass *Thalassia testudinum*. *Limnol. Oceanogr.* 15: 14–20.

Fenchel, T. 1974. Intrinsic rate of natural increase: the relationship with body size. *Oecologia (Berl.)* 14: 317–326.

Fenchel, T. 1975. The quantitative importance of the benthic microfauna of an arctic tundra pond. *Hydrobiologia* 46: 445–464.

Fenchel, T. 1980a. Suspension feeding in ciliated protozoa: structure and function of feeding organelles. *Arch. Protistenk.* 123: 239–260.

Fenchel, T. 1980b. Suspension feeding in ciliated protozoa: functional response and particle size selection. *Microb. Ecol.* 6: 1–11.

Fenchel, T. 1980c. Suspension feeding in ciliated protozoa: feeding rates and their ecological significance. *Microb. Ecol.* 6: 13–25.

Fenchel, T. 1980d. Relation between particle size selection and clearance in suspension feeding ciliates. *Limnol. Oceanogr.* 25: 733–738.

Fenchel, T. 1982a. Ecology of heterotrophic microflagellates. I. Some important forms and their functional morphology. *Mar. Ecol. Prog. Ser.* 8: 211–223.

Fenchel, T. 1982b. Ecology of heterotrophic microflagellates. II. Bioenergetics and growth. *Mar. Ecol. Prog. Ser.* 8: 225–231.

Fenchel, T. 1982c. Ecology of heterotrophic microflagellates. III. Adaptations to heterogeneous environments. *Mar. Ecol. Prog. Ser.* 9: 25–33.

Fenchel, T. 1982d. Ecology of heterotrophic flagellates. IV. Quantitative occurrence and importance as bacterial consumers. *Mar. Ecol. Prog. Ser.* 9: 35–42.

Fenchel, T. 1984. Suspended marine bacteria as food source. Pp. 301–315 in M. J. Fasham (ed.), *Energy and Materials in Marine Ecosystems.* Plenum Press, New York.

Fenchel, T. 1986a. Protozoan filter feeding. *Prog. Protistology*, 1 (in press).

Fenchel, T. 1986b. The ecology of heterotrophic flagellates. *Adv. Microb. Ecol.* 9: 57–97.

Fenchel, T. & Finlay, B. J. 1983. Respiration rates in heterotrophic, free-living Protozoa. *Microb. Ecol.* 9: 99–122.

Fenchel, T. & Finlay, B. J. 1984. Geotaxis in the ciliated protozoon *Loxodes. J. Exp. Biol.* 110: 17–33.

Fenchel, T. & Finlay, B. J. 1986a. The structure and function of Müller vesicles in loxodid ciliates. *J. Protozool.* 33: 69–79.

Fenchel, T. & Finlay, B. J. 1986b. The responses to light and to oxygen in the ciliated protozoon *Loxodes striatus. J. Protozool.* 33: 139–145.

Fenchel, T. & Harrison, P. 1976. The significance of bacterial grazing and mineral cycling for the decomposition of particulate detritus. Pp. 285–299 in J. M. Anderson & A. MacFadyen (eds.), *The Role of Terrestrial and Aquatic Organisms in Decomposition Processes.* Blackwell, Oxford.

Fenchel, T. & Patterson, D. J. 1986. *Percolomonas cosmopolitus* (Ruinen) n. gen., a new type of filter feeding flagellate from marine plankton. *J. Mar. Biol. Assoc. U.K.* 66: 465–482.

Fenchel, T., Perry, T. & Thane, A. 1977. Anaerobiosis and symbiosis with bacteria in free living ciliates. *J. Protozool.* 24: 154–163.

Fenchel, T. & Straarup, B. J. 1971. Vertical distribution of photosynthetic pigments and the penetration of light in marine sediments. *Oikos* 22: 172–182.

Finlay, B. J. 1978. Community production and respiration by ciliated protozoa in the benthos of a small eutrophic loch. *Freshwater Biol.* 8: 327–341.

Finlay, B. J. 1980. Temporal and vertical distribution of ciliophoran communities in the benthos of a small eutrophic loch with particular reference to the redox profile. *Freshwater Biol.* 10: 15–34.

Finlay, B. J. 1981. Oxygen availability and seasonal migrations of ciliated protozoa in a freshwater lake. *J. Gen. Microbiol.* 123: 173–178.

Finlay, B. J. 1982. Effects of seasonal anoxia on the community of benthic ciliated protozoa in a productive lake. *Arch. Protistenk.* 125: 215–222.

Finlay, B. J. 1985. Nitrate respiration by Protozoa (*Loxodes* spp.) in the hypolimnetic nitrite maximum of a productive freshwater pond. *Freshwater Biol.* 15: 333–346.

Finlay, B. J., Bannister, P. & Stewart, J. 1979. Temporal variation in benthic ciliates and the application of association analysis. *Freshwater Biol.* 9: 45–53.

Finlay, B. J. & Berninger, U.-G. 1984. Coexistence of congeric ciliates (Karyorelectida: *Loxodes*) in relation to food resources in two freshwater lakes. *J. Anim. Ecol.* 53: 929–943.

Finlay, B. J., Curds, C. R., Bamforth, S. S. & Bafort, J. J. 1986. Ciliated protozoa and other microorganisms from two African soda lakes (L. Nakuru and L. Simbi, Kenya). *Arch. Protistenk.* (in press).

Finlay, B. J. & Fenchel, T. 1986. Physiological ecology of the ciliated protozoon *Loxodes*. *Freshwater Biological Association Annual Report*, 54: 73–96. Ambleside, U.K.

Finlay, B. J., Fenchel, T. & Gardner, S. 1986. Oxygen perception and $O_2$ toxicity in the freshwater ciliated protozoon *Loxodes*. *J. Protozool.* 33: 157–165.

Finlay, B. J., Hetherington, N. B. & Davison, W. 1983. Active biological participation in lacustrine barium chemistry. *Geochim. Cosmochim. Acta.* 47: 1325–1329.

Finlay, B. J. & Ochsenbein-Gattlen, C. 1982. Ecology of Free-living Protozoa. A Bibliography. *Freshwater Biological Association*, Occasional Publication, No. 17. Ambleside, U.K.

Finlay, B. J., Span, A. S. W. & Harman, J. M. P. 1983a. Nitrate respiration in primitive eukaryotes. *Nature, Lond.* 303: 333–336.

Finlay, B. J., Span, A. & Ochsenbein-Gattlen, C. 1983b. Influence of physiological states on indices of respiration rate in protozoa. *Comp. Biochem. Physiol.* 74A: 211–219.

Finlay, B. J. & Uhlig, G. 1981. Calorific and carbon values of marine and freshwater Protozoa. *Helgoländer Meeresunters* 34: 401–412.

Finley, H. E. 1930. Toleration of freshwater Protozoa to increased salinity. *Ecology* 11: 337–346.

Foissner, W. 1979. Über ein massenauftreten von *Ophrydium eutrophicum* nov. spec. (Ciliophora, Peritrichida) und *Cristatella mucedo* Cuvier (Bryozoa, Cristalladidae) in zwei Voralpenseen (Wallersee, Fuschlsee). *Ber. Nat.-Med. Ver. Salzburg* 3/4: 95–100.

Foissner, W. 1982. Ökologie und Taxonomie der Hypotrichida (Protozoa: Ciliophora) einiger österreichicher Böden. *Arch. Protistenk.* 126: 19–143.

Foissner, W. 1985a. Klassifikation und Phylogenie der Colpodea (Protozoa: Ciliophora). *Arch. Protistenk.* 129: 239–290.

Foissner, W. 1985b. Morphologie und Infraciliatur der Genera *Microthorax* und *Stammeridium* und Klassifikation der Microthoracina Jankowski, 1967 (Protozoa: Ciliophora). *Zool. Anz. Jena* 214: 33–53.

Foissner, W. 1986. Soil protozoa: fundamental problems, ecological significance, adaptation, indicators of environmental quality, guide to the literature.
*Prog. Prostistology* 2 (in press).

Foissner, W. & Foissner, I. 1984. First record of an ectoparasitic flagellate on
ciliates: an ultrastructural investigation of the morphology and mode of attachment of *Spiromonas gonderi* nov. sp. (Zoomastigophora, Sarcomonadi-
dae) invading the pellicle of ciliates of the genus *Colpoda* (Ciliophora, Colpodida). *Protistologica* 20: 635–648.

Foissner, W. & Wilbert, N. 1979. Morphologie, Infraciliatur und ökologie der
limnischen Tintinnina: *Tintinnidium fluviatile* Stein, *Tintinnidium pusillum*
Entz, *Tintinnopsis cylindrata* Daday und *Codonella cratera* (Leidy) (Ciliophora, Polyhymenophora). *J. Protozool.* 26: 90–103.

Foster, K. W. & Smyth, R. D. 1980. Light antennas in phototactic algae. *Microbiol. Rev.* 44: 572–630.

Fraenkel, G. & Gunn, D. 1940. *The Orientation of Animals.* Oxford University
Press, London and New York.

Gabel, B. 1971. Die Foraminiferen der Nordsee. *Helgoländ. Wiss. Meeresunters.*
22: 1–65.

Gaines, G. & Taylor, F. J. R. 1985. Form and function of the dinoflagellate
transverse flagellum. *J. Protozool.* 32: 290–296.

Gause, G. F. 1934. *The Struggle for Existence.* Williams and Wilkins, Baltimore.

Gerlach, S. A., Hahn, A. E. & Schrage, M. 1985. Size spectra of benthic biomass
and metabolism. *Mar. Ecol. Prog. Ser.* 26: 161–173.

Giese, A. C. 1973. *Blepharisma.* Stanford University Press, Stanford, California.

Glaessner, M. F. 1984. *The Dawn of Animal Life.* Cambridge University Press,
Cambridge.

Glibert, P. M. 1982. Regional studies of daily, seasonal and size fraction variability in ammonium regeneration. *Mar. Biol.* 70: 209–222.

Goldman, J. C. 1984. Conceptual role for microaggregates in pelagic waters.
*Bull. Mar. Sci.* 35: 462–476.

Goldman, J. C., Caron, D. A., Andersen, O. K. & Dennett, M. R. 1985. Nutrient
cycling in a microflagellate food chain: I. Nitrogen dynamics. *Mar. Ecol. Prog.
Ser.* 24: 231–242.

Golemansky, V. 1978. Adaptations morphologiques des thécamoebiens psam-
mobiontes du psammal supralittoral des mer. *Acta Protozool.* 17: 141–152.

Golemansky, V. & Ogden, C. 1980. Shell structure of three littoral species of
testate amoebae from the Black Sea (Rhizopodea: Protozoa). *Bull. Br. Mus:
Nat. Hist. (Zool.)* 38: 1–6.

Goulder, R. 1974. The seasonal and spatial distribution of some benthic ciliated
Protozoa in Esthwaite Water. *Freshwater Biol.* 4: 127–147.

Grassé, P.-P. 1952. *Traité de Zoologie, 1. Protozoaires.* Masson et Cⁱᵉ, Paris.

Grell, K. G. 1973. *Protozoology.* Springer-Verlag, Berlin and New York.

Greuet, C. 1968. Organisation ultrastructurale de l'ocelle de deux peridiniens
Warnowiidae, *Erythropsis pavillardi* Kofoid et Swezy et *Warnowia pulchra*
Schiffer. *Protistologica* 4: 209–230.

Greuet, C. 1969. Étude morphologique et ultrastructurale du trophonte d'*Erythropsis pavillardi* Kofoid et Swezy. *Protistologica* 5: 481–503.

Griessmann, K. 1914. Über marine Flagellaten. *Arch. Protistenk.* 32: 1–78.

Grolière, C.-A. & Njiné, T. 1973. Étude comparée de la dynamique des populations de ciliés dans differents biotypes d'une mare de forêt pendant une année. *Protistologica* 9: 5–16.

Habte, M. & Alexander, M. 1978. Protozoan density and the coexistence of protozoan predators and bacterial prey. *Ecology* 59: 140–146.

Haeckel, E. 1866. *Generelle Morphologie der Organismen.* 2. vols. G. Reimer, Berlin.

Haeckel, E. 1887. Report on the scientific results of the voyage of H.M.S. Challenger. *Zoology* 18: 1–1760. Her Majesty's Stationary Office, Edinburgh.

Hänel, K. 1979. Systematik und ökologie der farblosen Flagellaten des Abwassers. *Arch. Protistenk.* 121: 73–137.

Hanson, E. D. 1976. Major evolutionary trends in animal protists. *J. Protozool.* 23: 4–12.

Hargraves, P. E. 1981. Seasonal variations of tintinnids (Ciliophora: Oligotrichida) in Narragansett Bay, Rhode Island, U.S.A. *J. Plankton Res.* 3: 81–91.

Harrison, W. G. 1978. Experimental measurements of nitrogen remineralization in coastal waters. *Limnol. Oceanogr.* 23: 684–694.

Hartwig, E. 1973. Die Ciliaten des Gezeiten-Sandtrandes der Nordseeinsel Sylt. I+II. *Mikrofauna des Meeresbodens* 18: 1–69 and 23: 1–71.

Hartwig, E. 1980. A bibliography of the interstitial ciliates (Protozoa): 1926–1979. *Arch. Protistenk.* 123: 422–438.

Hausmann, K. & Patterson, D. J. 1982. Pseudopod formation and membrane production during prey capture by a heliozoon (feeding by *Actinophrys*, II). *Cell Motility* 2: 9–24.

Heal, O. W. 1961. The distribution of testate amoebae (Rhizopoda, Testacea) in some fens and bogs in Northern England. *J. Linn. Soc. (Zool.)* 44: 369–382.

Heal, O. W. 1962. The abundance and micro-distribution of testate amoebae (Rhizopoda: Testacea) in *Sphagnum. Oikos* 13: 35–47.

Heal, O. W. 1964a. Observations on the seasonal and spatial distribution of Testacea (Protozoa: Rhizopoda) in *Sphagnum. J. Anim. Ecol.* 33: 395–412.

Heal, O. W. 1964b. The use of cultures for studying Testacea (Protozoa, Rhizopoda) in soil. *Protistologica* 4: 1–7.

Hecky, R. E. & Kling, H. J. 1981. The phytoplankton and protozooplankton of the euphotic zone of Lake Tanganyika: species composition, biomass, chlorophyll content, and spatio-temporal distribution. *Limnol. Oceanogr.* 26: 548–564.

Hegner, R. W. 1938. *Big Fleas Have Little Fleas or Who's Who Among the Protozoa.* Williams & Wilkins, Baltimore. (Reprinted 1968 by Dover, New York.)

Heinbokel, J. F. 1978. Studies on the functional role of tintinnids in the Southern California Bight. I. Grazing and growth rates in laboratory cultures. *Mar. Biol.* 47: 177–189.

Heinbokel, J. F. & Beers, J. R. 1979. Studies on the functional role of tintinnids in the southern California Bight. III. Grazing impact of natural assemblages. *Mar. Biol.* 52: 23–32.

Hemmingsen, A. M. 1960. Energy metabolism as related to body size and respiratory surfaces and its evolution. *Rep. Steno. Mem. Hosp. Copenhagen* 9: 1–110.

Hewett, S. W. 1980. Prey-dependent cell size in a protozoan predator. *J. Protozool.* 27: 311–313.

Hibberd, D. J. 1977. Observations on the ultrastructure of the cryptomonad endosymbiont of the red-water ciliate *Mesodinium rubrum. J. Mar. Biol. Assoc., U.K.* 57: 45–61.

Höglund, H. 1947. Foraminifera in the Gullmar Fjord and the Skagerak. *Zool. Bidrag Uppsala* 26: 1–328.

Holwill, M. E. J. 1974. Hydrodynamic aspects of ciliary and flagellar movement. Pp. 143–175 in M. A. Sleigh (ed.), *Cilia and Flagella.* Academic Press, London.

Hungate, R. E. 1955. Mutualistic intestinal protozoa. Pp. 159–199 in S. H. Hutner & A. Lwoff (eds.), *Biochemistry and Physiology of Protozoa.* Academic Press, New York.

Hungate, R. E. 1975. The rumen microbial ecosystem. *Ann. Rev. Ecol. Syst.* 6: 39–66.

Hutner, S. H. 1975. Maintaining protozoa and protozoan diversity in a culture collection. Pp. 43–52 in R. R. Colwell (ed.), *The Role of Culture Collections in the Era of Molecular Biology.* Am. Soc. Microbiol., Washington, D.C.

Ibanez, F. & Rassoulzadegan, F. 1977. A study of the relationships between pelagic ciliates (Oligotrichina) and planktonic nanoflagellates of the neritic ecosystem of the bay of Villefranche-sur-Mer. Analysis of chronological series. *Ann. Inst. Océanogr., Paris* 53: 17–30.

Issel, R. 1910. La faune de source thermales de Viterbo. *Int. Rev. Ges. Hydrobiol. Hydrogr.* 3: 178–180.

Iwatsuki, K. & Naitoh, Y. 1983. Behavioral responses in *Paramecium multimicronucleatum* to visible light. *Photochem. Photobiol.* 37: 415–419.

Jahn, T. L., Bovee, E. C. & Jahn, F. F. 1979. *How to know the Protozoa.* Wm. C. Brown Company Publishers, Dubuque, Iowa.

Jankowski, A. W. 1963. Pathology of Ciliophora. II. Life cycles of Suctoria parasiting in *Urostyla* and *Paramecium* (in Russian). *Tsitologia* 5: 428–439.

Jeon, K. W. 1983. Integration of bacterial endosymbionts in amoebae. *Int. Rev. Cytol.,* Suppl. 14: 29–47.

Johannes, R. E. 1965. Influence of marine protozoa on nutrient regeneration. *Limnol. Oceanogr.* 10: 434–442.

Jørgensen, C. B. 1976. August Pütter, August Krogh, and modern ideas on the use of dissolved organic matter in aquatic environments. *Biol. Rev.* 51: 291–328.

Kahl, A. 1928. Die Infusorien (Ciliata) der Oldesloer Salzwasserstellen. *Arch. Hydrobiol.* 19: 50–123, 189–246.

Kahl, A. 1933. Ciliata libera et ectocommensalia. Pp. 29–146 in G. Grimpe & E. Wagler (eds.) *Die Tierelt der Nord—und Ostsee.* Lief 23 (Teil II,C₃), Leipzig.

Karakashian, M. W. 1975. Symbiosis in *Paramecium bursaria. Symp. Soc. Exp. Biol.* 29: 145–173.

Kerr, S. R. 1974. Theory of size distribution in ecological communities. *J. Fish. Res. Board.* Canada 31: 1859–1862.

King, K. R., Hollibaugh, J. T. & Azam, F. 1980. Predator-prey interactions between the larvacean *Oikopleura dioica* and bacterioplankton in enclosed water columns. *Mar. Biol.* 56: 49–57.

Kirby, H. 1934. Some ciliates from salt marshes in California. *Arch. Protistenk.* 82: 114–133.

Koch, A. L. 1971. The adaptive responses of *Escherichia coli* to a feast and famine existence. *Adv. Microb. Physiol.* 6: 147–217.

Kolkwitz, R. & Marsson, M. 1909. Ökologie der tierischen Saprobien. *Int. Revue Ges. Hydrobiol.* 2: 1–126.

Kremer, B., Schmaljohann, R. & Röttger, R. 1980. Features and nutritional significance of photosynthates produced by unicellular algae symbiotic with large foraminifera. *Mar. Ecol. Prog. Ser.* 2: 225–228.

Kuhlmann, H.-W. & Heckmann, K. 1985. Interspecific morphogens regulating prey-predator relationships in protozoa. *Science* 227: 1347–1349.

Kuhlmann, S., Patterson, D. J. & Hausmann, K. 1980. Untersuchungen zu Nahrungserwerts und Nahrungsaufname bei *Homalozoon vermiculare*, Stokes 1887. *Protistologica* 16: 39–55.

Lackey, J. B. 1938a. A study of some ecologic factors affecting the distribution of Protozoa. *Ecol. Monogr.* 8: 501–527.

Lackey, J. B. 1938b. The fauna and flora of surface waters polluted by acid mine drainage. *U.S. Pub. Hlth. Rep.* 53: 1499–1507.

Lapidus, R. & Levandowsky, M. 1981. Mathematical models of behavioral responses to sensory stimuli by Protozoa. Pp. 235–260 in M. Levandowsky & S. H. Hutner (eds.), *Biochemistry and Physiology of Protozoa*. 2nd ed., vol. 4. Academic Press, New York.

Laval, M. 1971. Ultrastructure et mode de nutrition du choanoflagellé *Salpinoeca pelagica* sp. nov. Comparaison avec les choanocytes des spongiaires. *Protistologica* 7: 325–336.

Laval-Peuto, M. 1981. Construction of the lorica in ciliata Tintinnina. In vivo study of *Favella ehrenbergi*: variability of the phenotypes during the cycle, biology, statistics, biometry. *Protistologica* 17: 249–272.

Leadbeater, B. S. C. & Morton, C. 1974. A microscopical study of a marine species of *Codosiga* James-Clark (Choanoflagellata) with special reference to the ingestion of bacteria. *Biol. J. Linn. Soc.* 6: 337–347.

Lee, C. C. & Fenchel, T. 1972. Studies on ciliates associated with sea ice from Antarctica. II. Temperature responses and tolerances in ciliates from Antarctic, temperate and tropical habitats. *Arch. Protistenk.* 114: 237–244.

Lee, J. J. 1980. Nutrition and physiology of the Foraminifera. Pp. 43–66 in M. Levandowsky & S. H. Hutner (eds.), *The Biochemistry and Physiology of Protozoa*. 2nd ed., vol. 3. Academic Press, New York.

Lee, J. J. 1983. Perspective on algal endosymbiosis in larger foraminifera. *Int. Rev. Cytol.*, suppl. 14: 49–77.

Lee, J. J. & Corliss, J. O. (eds.) 1985. Symposium on "Symbiosis in Protozoa." *J. Protozool.* 32: 371–403.

Lee, J. J., Hutner, S. H. & Bovee, E. C. (eds.) 1985. *An Illustrated Guide to the Protozoa*. Society of Protozoologists, P.O. Box 368, Lawrence, Kansas.

Lee, J. J. & McEnery, M. E. 1983. Symbiosis in foraminifera. Pp. 37–68 in L. J. Goff (ed.), *Algal Symbiosis*. Cambridge University Press, Cambridge.

Lee, J. J., McEnery, M. E. & Rubin, H. 1969. Quantitative studies on the growth of *Allogromia laticollaris* (Foraminifera). *J. Protozool.* 16: 377–395.

Lee, J. J. & Muller, W. A. 1973. Trophic dynamics and niches of salt marsh Foraminifera. *Am. Zool.* 13: 215–223.

Legner, M. 1973. Experimental approach to the role of protozoa in aquatic ecosystems. *Amer. Zool.* 13: 177–192.

Legner, M. 1975. Concentration of organic substances in water as a factor controlling the occurrence of some ciliate species. *Int. Revue Ges. Hydrobiol.* 60: 639–654.

Lessard, E. J. & Swift, E. 1985. Species-specific grazing rates of heterotrophic dinoflagellates in oceanic waters measured with a dual-label radioisotope technique. *Mar. Biol.* 87: 289–296.

Levandowsky, M., Cheng, T., Kehr, A., Kim, L., Gardner, L., Silvern, L., Tsang, L., Lai, G., Chung, C. & Prakash, E. 1984. Chemosensory responses to amino acids and certain amines by the ciliate *Tetrahymena*: a flat capillary assay. *Biol. Bull.* 167: 322–330.

Levandowsky, M. & Hutner, S. H. (eds.) 1978–81. *Biochemistry and Physiology of Protozoa.* 2nd ed. Vol. 1 (1979), vol. 2 (1979), vol. 3 (1980), vol. 4 (1981). Academic Press, New York.

Levin, S. A. & Paine, R. T. (1974) Disturbance, patch formation and community structure. *Proc. Nat. Acad. Sci. U.S.A.* 71: 2744–2747.

Levin, S. A. & Segel, L. A. 1985. Pattern generation in space and aspect. *SIAM Rev.* 27: 45–67.

Levins, R. 1979. Coexistence in a variable environment. *Am. Nat.* 114: 765–783.

Lighthart, B. 1969. Planktonic and benthic bacterivorous protozoa at eleven stations in Puget Sound and adjacent Pacific Ocean. *J. Fish Res. Bd. Canada* 26: 299–304.

Lighthill, J. 1976. Flagellar hydrodynamics. *SIAM Rev.* 18: 161–230.

Lindberg, R. E. & Bovee, E. C. 1976. *Chaos carolinensis*, induction of phagocytosis and cannibalism. *J. Protozool.* 23: 333–336.

Linley, E. A. S., Newell, R. C. & Bosman, S. A. 1981. Heterotrophic utilization of mucilage released during fragmentation of kelp (*Ecklonia maxima* and *Laminaria pallida*). I. Development of microbial communities associated with the degradation of kelp mucilage. *Mar. Ecol. Prog. Ser.* 4: 31–41.

Linnenbach, M., Hausmann, K. & Patterson, D. J. 1983. Ultrastructural studies on the food vacuole cycle of a heliozoon (Feeding by *Actinophrys*, III). *Protoplasma* 115: 43–51.

Lipps, J. H. & Valentine, J. W. 1970. The role of foraminifera in the trophic structure of marine communities. *Lethaia* 3: 279–286.

Lohmann, H. 1911. Über das Nannoplankton und die Zentrifugierung kleinster Wasserproben zur Gewinnung desselben in lebendem Zustande. *Int. Rev. Ges. Hydrobiol. Hydrogr.* 4: 1–38.

Lopez, E. 1979. Algal chloroplasts in the protoplasm of three species of benthic foraminifera: taxonomic affinity, viability and persistence. *Mar. Biol.* 53: 201–211.

Lousier, J. D. 1985. Population dynamics and production studies of species of Centropyxidae (Testacea, Rhizopoda) in an aspen woodland soil. *Arch. Protistenk.* 130: 165–178.

Lousier, J. D. & Parkinson, D. 1984. Annual population dynamics and production ecology of testacea (Protozoa, Rhizopoda) in an aspen woodland soil. *Soil. Biol. Biochem.* 16: 103–114.

Luckinbill, L. S. 1973. Coexistence in laboratory populations of *Paramecium aurelia* and its predator *Didinium nasutum*. *Ecology* 54: 1320–1327.

Luckinbill, L. S. 1974. The effects of space and enrichment on a predator-prey system. *Ecology* 55: 1142–1147.

Luckinbill, L. S. 1979. Selection and the r/K continuum in experimental populations of protozoa. *Am. Nat.* 113: 427–437.

Lüttenegger, G., Foissner, W. & Adam, H. 1985. r—and K-selection in soil ciliates: a field and experimental approach. *Oecologia (Berl.)* 66: 574–579.

Lutze, G. 1965. Zur Foraminiferen-Fauna der Ostsee. *Meyniana Kiel* 15: 75–142.

MacArthur, R. H. & Levins, R. 1967. The limiting similarity, convergence and divergence of coexisting species. *Am. Nat.* 101: 377–385.

MacArthur, R. H. & Wilson, E. O. 1967. *The Theory of Island Biogeography*. Princeton University Press, Princeton.

Machemer, H. 1974. Ciliary activity and metachronism in protozoa. Pp. 199–286 in M. A. Sleigh (ed.), *Cilia and Flagella*. Academic Press, London and New York.

Madoni, P. & Ghetti, P. F. 1981. The structure of ciliated protozoa communities in biological sewage-treatment plants. *Hydrobiologia* 83: 207–215.

Maguire, B. Jr. 1963. The passive dispersal of small aquatic organisms and their colonization of isolated bodies of water. *Ecol. Monogr.* 33: 161–185.

Maguire, B. Jr. 1977. Community structure of protozoans and algae with particular emphasis on recently colonized bodies of water. Pp. 355–397 in J. Cairns Jr. (ed.), *Aquatic Microbial Communities*. Garland Publishing, Inc., New York and London.

Mare, M. F. 1942. A study of a marine benthic community with special reference to the microorganisms. *J. Mar. Biol. Assoc. U.K.* 25: 517–554.

Margulis, L. 1981. *Symbiosis in Cell Evolution*. W. H. Freeman and Company, San Francisco.

Matera, N. J. & Lee, J. J. 1972. Environmental factors affecting the standing crop of foraminifera in sublittoral and psammolittoral communities of a Long Island salt marsh. *Mar. Biol.* 14: 89–103.

Matsuoka, T. 1983. Negative phototaxis in *Blepharisma japonicum*. *J. Protozool.* 30: 409–414.

May, R. M. 1972. Limit cycles in predator-prey communities. *Science* 177: 900–902.

May, R. M. 1973. *Stability and Complexity in Model Ecosystems*. Princeton University Press, Princeton.

May, R. M. 1978. The evolution of ecological systems. *Sci. Amer.* 239,3: 118–133.

Maynard Smith, J. 1978. *The Evolution of Sex*. Cambridge University Press, Cambridge.

McMahon, T. A. & Bonner, J. T. 1983. *On Size and Life*. W. H. Freeman and Company, New York.

Meisterfeld, R. 1977. Die horizontale und vertikale Verteilung der Testaceen (Rhizopoda, Testacea) in *Sphagnum*. *Arch. Hydrobiol.* 79: 319–356.

Mignot, J.-P. 1966. Structure et ultrastructure de quelques Euglenomonadines. *Protistologica* 2: 51–117.

Miller, S. & Diehn, B. 1978. Cytochrome c oxidase as the receptor molecule for chemo-accumulation (chemotaxis) of *Euglena* towards oxygen. *Science* 200: 548–549.

Mitchell, J. G., Okubo, A. & Fuhrman, J. A. 1985. Microzones surrounding phytoplankton forms the basis for a stratified marine microbial ecosystem. *Nature, Lond.* 316: 58–59.

Miyake, A. 1981. Physiology and biochemistry of conjugation in ciliates. Pp. 125–198 in M. Levandowsky & S. H. Hutner (eds.), *Biochemistry and Physiology of Protozoa*. 2nd ed., vol. 4. Academic Press, New York.

Mueller, J. A. & Mueller, W. P. 1970. *Colpoda culcullus*: a terrestrial aquatic. *Amer. Midl. Nat.* 84: 1–12.

Mueller, M., Röhlich, P. & Törö, I. 1965. Studies on feeding and digestion in protozoa. VII. Ingestion of polystyrene latex particles and its early effect on acid phosphatase in *Paramecium multimicronucleatum* and *Tetrahymena pyriformis*. *J. Protozool.* 12: 27–34.

Müller, M. 1980. The hydrogenosome. *Soc. Gen. Microb. Symposium* 30: 127–142.

Muller, P. 1978. Carbon fixation and loss in a foraminiferal-algal symbiont system. *J. Foram. Res.* 8: 35–41.

Muller, W. A. 1975. Competition for food and other niche-related studies of three species of salt-marsh foraminifera. *Mar. Biol.* 31: 339–351.

Muscatine, L. & Poole, R. R. 1979. Regulation of numbers of intracellular algae. *Proc. R. Soc. London.* B. 204: 131–139.

Naitoh, Y. & Eckert, R. 1974. The control of ciliary activity in Protozoa. Pp. 305–352 in M. A. Sleigh (ed.), *Cilia and Flagella*. Academic Press, London.

Naitoh, Y. & Sugino, K. 1984. Ciliary movement and its control in *Paramecium J. Protozool.* 31: 31–40.

Nanney, D. L. 1982. Genes and phenes in *Tetrahymena*. *BioScience* 32: 783–788.

Nanney, D. L. 1985. The tangled tempos underlying *Tetrahymena* taxonomy. *Atti Soc. Tosc. Sci. Nat., Mem.*, Ser. B, 92: 1–13.

Nanney, D. L., Cooper, L. E., Simon, E. M. & Whitt, G. S. 1980. Isozymic characterization of three mating groups of the *Tetrahymena pyriformis* complex. *J. Protozool.* 27: 451–459.

Nelsen, E. M. & DeBault, L. 1978. Transformation in *Tetrahymena pyriformis*: description of an inducible phenotype. *J. Protozool.* 25: 113–119.

Nenninger, U. 1948. Die Peritrichen der Umgebung von Erlangen mit besonderer Berücksichtigung ihrer Wirtsspecifität. *Zool. Jb., Abt. Syst.* 77: 169–266.

Newell, R. C. & Linley, E. A. S. 1984. Significance of microheterotrophs in the decomposition of phytoplankton: estimates of carbon and nitrogen flow based on the biomass of plankton communities. *Mar. Ecol. Prog. Ser.* 16: 105–119.

Nilsson, J. R. 1970. Cytolosomes in *Tetrahymena pyriformis* GL. *C. R. Trav. Lab. Carlsberg* 38: 87–121.

Nilsson, J. R. 1979. Phagotrophy in *Tetrahymena*. Pp. 339–379 in M. Levandowsky & S. H. Hutner (eds.), *Biochemistry and Physiology of Protozoa.* 2nd ed., vol. 2. Academic Press, New York.

Nisbet, B. 1974. An ultrastructural study of the feeding apparatus of *Peranema trichophorum. J. Protozool.* 21: 39–48.

Nisbet, B. 1984. *Nutrition and Feeding Strategies in Protozoa.* Croom Helm, London.

Nobili, R., Luporini, P. & Dini, F. 1978. Breeding systems, species relationships and evolutionary trends in some marine species of Euplotidae (Hypotrichida Ciliata). Pp. 591–616 in B. Battaglia & J. A. Beardmore (eds.), *Marine Organisms: Genetics, Ecology, and Evolution.* Plenum, New York.

Noland, L. E. & Gojdics, M. 1967. Ecology of free-living protozoa. Pp. 215–266 in Tze-Tuan Chen (ed.), *Research in Protozoology.* Vol. 2. Pergamon Press, Oxford.

Nyberg, D. 1974. Breeding systems and resistance to environmental stress in ciliates. *Evolution* 28: 367–380.

Nyholm, K.-G. 1950. A marine nude rhizopod type *Megamoebamyxa argillobia. Zool. Bidrag. Uppsala* 29: 93–102.

Nyholm, K.-G. & Gertz, I. 1973. To the biology of the monothalamous foraminifer *Allogromia marina* n. sp. *Zoon* 1: 89–93.

Oakley, B. R. & Taylor, F. J. R. 1978. Evidence for a new type of endosymbiotic organization in a population of the ciliate *Mesodinium rubrum* from British Columbia. *BioSystems* 10: 361–369.

Ogden, C. G. & Hedley, R. H. 1980. *An Atlas of Freshwater Testate Amoebae.* Oxford University Press, Oxford.

Okubo, A. 1980. *Diffusion and Ecological Problems: Mathematical Models.* Springer-Verlag, Berlin.

Old, K. M. & Darbyshire, J. F. 1980. *Arachnula impatiens* Cienk., a mycophagous giant amoeba from soil. *Protistologica* 16: 277–287.

Olive, L. S. & Blanton, R. L. 1980. Aerial sorocarp development by the aggregative ciliate *Sorogena stoianovitchae. J. Protozool.* 27: 293–299.

Paasche, E. & Kristiansen, S. 1982. Ammonium regeneration by micro-zooplankton in the Oslofjord. *Mar. Biol.* 69: 55–63.

Pace, M. L. 1982. Planktonic ciliates: their distribution, abundance, and relationship to microbial resources in a monomictic lake. *Can. J. Fish. Aquat. Sci.* 39: 1106–1116.

Pace, M. L. & Orcutt, J. D. Jr. 1981. The relative importance of protozoans, rotifers, and crustaceans in a freshwater zooplankton community. *Limnol. Oceanogr.* 26: 822–830.

Pack, A. 1919. Two ciliata of the Great Salt Lake. *Biol. Bull.* 36: 273–282.

Packard, T. T., Blasco, D. & Barber, R. T. 1978. *Mesodinium rubrum* in Baja California upwelling system. Pp. 73–89 in R. Boje & M. Tomczak (eds.), *Upwelling Systems.* Springer-Verlag, Berlin.

Page, F. C. 1974. Some marine *Platyamoeba* of East Anglia. *J. Mar. Biol. Assoc. U.K.* 54: 651–664.

Page, F. C. 1976. *An illustrated key to Freshwater and Soil Amoebae.* Freshwater Biological Association, Ambleside, U.K.

Page, F. C. 1983. *Marine Gymnamoebae.* Institute of Terrestrial Ecology, NERC, Cambridge, U.K.

Paine, W. J. 1970. Energy yields and growth of heterotrophs. *Ann. Rev. Microbiol.* 24: 17–52.

Patterson, D. J. 1980. Contractile vacuoles and associated structure: their organization and function. *Biol. Rev.* 55: 1–46.

Patterson, D. J. & Fenchel, T. 1985. Insights into the evolution of heliozoa (Protozoa, Sarcodina) as provided by ultrastructural studies on a new species of flagellate from the genus *Pteridomonas. Biol. J. Limn. Soc.* 34: 381–403.

Pavlovskaya, T. V. & Pechen, G. A. 1971. Infusoria as a food of some mass species of marine planktonic animals (in Russian). *Zool. Zh.* 50: 633–640.

Peck, R. K. 1985. Feeding behavior in the ciliate *Pseudomicrothorax dubius* in a series of morphologically and physiologically distinct events. *J. Protozool.* 32: 492–501.

Pedrós-Alió, C. & Brock, T. D. 1983. The impact of zooplankton feeding on the epilimnetic bacteria of a eutrophic lake. *Freshwater Biol.* 13: 227–239.

Persoone, G. 1968. Écologie des infusoires dans les salissures de substrats immergés dans un port de mer. *Protistologica* 4: 187–194.

Peterson, B. J., Hobbie, J. E. & Haney, J. F. 1978. Daphnia grazing on natural bacteria. *Limnol. Oceanogr.* 23: 1039–1044.

Petz, W., Foissner, W. & Adam, H. 1985. Culture, food selection and growth rate in the mycophagous ciliate *Grossglockneria acuta* Foissner, 1980: first evidence of autochthonous soil ciliates. *Soil. Biol. Biochem.* 17: 871–875.

Pianka, E. R. 1978. *Evolutionary Ecology.* 2nd ed. Harper & Row, New York.

Picken, L. E. R. 1937. The structure of some protozoan communities. *J. Ecol.* 25: 368–384.

Pill-Soon, S. & Walker, E. B. 1981. Molecular aspects of photoreceptors in protozoa and other microorganisms. Pp. 199–233 in M. Levandowsky & S. H. Hutner (eds.), *Biochemistry and Physiology of Protozoa.* 2nd ed. Vol. 4. Academic Press, New York.

Platt, T. & Denman, K. 1977. Organisation in the pelagic ecosystem. *Helgoländer Wiss. Meeresunters.* 30: 575–581.

Prasad, R. R. 1958. A note on the occurrence and feeding habit of *Noctiluca* and their effects on the plankton communities and fisheries. *Proc. Ind. Acad. Sci.* B, 47: 331–337.

Precht, H. 1935. Epizoen der Kieler Bucht. *Nova Acta Leopoldina* 3: 405–474.

Preer, J. R. Jr., Preer, L. B. & Jurand, A. 1974. Kappa and other endosymbionts in *Paramecium aurelia. Bact. Rev.* 38: 113–163.

Purcell, E. M. 1977. Life at low Reynolds number. *Amer. J. Phys.* 45: 3–11.

Raikov, I. B. 1982. *The Protozoan Nucleus.* Springer-Verlag, Wien, New York.

Rasmussen, L. & Orias, E. 1975. *Tetrahymena*: growth without phagocyosis. *Science* 190: 464–465.

Rassoulzadegan, F. 1977. Évolution anuelle des ciliés pélagiques en Méditerranée nord-occidentale ciliés oligotriches "non tintinnides" (Oligotrichina). *Ann. Inst. Océanogr. Paris* 53: 125–134.

Rassoulzadegan, F. 1982. Dependence of grazing rate, gross growth efficiency, and food size range on temperature in a pelagic, oligotrichous ciliate *Lohmanniella spiralis* Leeg., fed on naturally occurring particulate matter. *Ann. Inst. Océanogr., Paris* 58: 177–184.

Rassoulzadegan, F. & Etienne, M. 1981. Grazing rate of the tintinnid *Stenosomella ventricosa* (Clap. & Lachm.) Jörg. on the spectrum of the naturally occurring particulate matter from a Mediterranean neritic area. *Limnol. Oceanogr.* 26: 258–270.

Revill, D. L., Stewart, K. W. & Schlichting, H. E. Jr. 1967. Passive dispersal of viable algae and protozoa by certain craneflies and midges. *Ecology* 48: 1023–1027.

Riemann, B. 1985. Potential importance of fish predation and zooplankton grazing on natural populations of freshwater bacteria. *Appl. Environ. Microbiol.* 50: 187–193.

Roberts, A. M. 1970. Geotaxis in motile microorganisms. *J. Exp. Biol.* 53: 687–699.

Roberts, A. M. 1981. Hydrodynamics in protozoan swimming. Pp. 6–66 in M. Levandowsky & S. H. Hutner (eds.), *Biochemistry and Physiology of Protozoa*, 2nd ed., vol. 4. Academic Press, New York.

Rosenzweig, M. L. 1971. Paradox of enrichment: destabilization of exploitation ecosystems in ecological time. *Science* 171: 385–387.

Rottgardt, D. 1952. Mikropaläontologisch wichtige Bestandteile rezenter brackischer Sedimente an der Küsten Schleswig-Holsteins. *Meyniana, Kiel* 1: 169–228.

Röttger, R. 1972. Die Kultur von *Heterostegina depressa* (Foraminifera: Nummultidae). *Mar. Biol.* 15: 150–159.

Röttger, R. & Berger, W. H. 1972. Benthic foraminifera: morphology and growth in clone cultures of *Heterostegina depressa*. *Mar. Biol.* 15: 89–94.

Rubinow, S. I. 1975. *Introduction to Mathematical Biology*. John Wiley & Sons, New York.

Russel, E. W. 1973. *Soil Conditions and Plant Growth*. 10th ed. Longman, London.

Sahling, G. & Uhlig, G. 1982. Rhythms and distributional phenomena in *Noctiluca miliaris*. *Ann. Inst. Océnogr., Paris* 58: 277–284.

Salt, G. W. 1979. Density, starvation, and swimming rate in *Didinium* populations. *Am. Nat.* 113: 135–143.

Satir, P. 1984. The generation of ciliary motion. *J. Protozool.* 31: 8–12.

Savoie, A. 1968. Les ciliés histophages en biologie cellulaire. *Ann. Univ. Ferrara, NS* 3,6: 65–71.

Schlichting, H. E. Jr. & Sides, S. L. 1969. The passive transport of aquatic microorganisms by selected Hemiptera. *J. Ecol.* 57: 759–764.

Schmidt-Nielsen, K. 1984. *Scaling. Why is Animal Size so Important?* Cambridge University Press, Cambridge.

Schönborn, W. 1962. Über Planktismus und Zyklomorphose bei *Difflugia limnetica* (Levander) Penard. *Limnologica (Berl.)* 1: 21–34.

Schönborn, W. 1977. Production studies on Protozoa. *Oecologica (Berl.)* 27: 171–184.

Schönborn, W. 1982. Estimation of annual production of testacea (Protozoa) in mull and moder (II). *Pedobiologia* 23: 383–393.

Schuster, F. L. 1979. Small amoebas and amoeba flagellates. Pp. 215–285 in M. Levandowsky & S. H. Hutner (eds.), *Biochemistry and Physiology of Protozoa.* 2nd ed., vol. 1. Academic Press, New York.

Sheldon, R. W., Prakash, A. & Sutcliffe, W. H. Jr. 1972. The size distribution of particles in the ocean. *Limnol. Oceanogr.* 17: 327–340.

Sherr, B. F., Sherr, E. B. & Berman, T. 1982. Decomposition of organic detritus: a selective role for microflagellate protozoa. *Limnol. Oceanogr.* 27: 765–769.

Sherr, B. F., Sherr, E. B. & Berman, T. 1983. Grazing, growth and ammonium excretion rates of a heterotrophic microflagellate fed with four species of bacteria. *Appl. Environ. Microbiol.* 45: 1196–1201.

Sherr, B. F., Sherr, E. B. & Newell, S. Y. 1984. Abundance and productivity of heterotrophic nanoplankton in Georgia coastal waters. *J. Plank. Res.* 6: 195–203.

Sherr, E. B., Sherr, B. F., Fallon, R. D. & Newell, S. Y. 1986. Small aloricate ciliates as a major component of the marine heterotrophic nanoplankton. *Limnol. Oceanogr.* 31: 177–183.

Sieburth, J. McN. 1979. *Sea Microbes.* Oxford University Press, New York.

Sieburth, J. McN. & Davis, P. G. 1982. The role of heterotrophic nanoplankton in the grazing and nurturing of planktonic bacteria in the Sargasso and Caribbean Seas. *Ann. Inst. Oceanogr., Paris* 58: 285–296.

Silver, M. W., Gowing, M. M., Brownlee, D. C. & Corliss, J. O. 1984. Ciliated protozoa associated with oceanic sinking detritus. *Nature, Lond.* 309: 246–248.

Silvester, N. R. & Sleigh, M. A. 1985. The forces on microorganisms at surfaces in flowing water. *Freshwater Biol.* 15: 433–448.

Skuja, H. 1956. Taxonomische un biologische Studien über das Phytoplankton schwedischer Binnengewässer. *Nova Acta Regia Soc. Sci. Upsaliensis*, Ser 4, 16: 1–404.

Sleigh, M. A. (ed.) 1974. *Cilia and Flagella.* Academic Press, London.

Sleigh, M. A. 1984. The integrated activity of cilia: function and coordination. *J. Protozool.* 31: 16–21.

Sleigh, M. A. & Blake, J. R. 1977. Methods of ciliary propulsion and their size limitations. Pp. 243–256 in T. J. Pedley (ed.), *Scale Effects in Animal Locomotion.* Academic Press, London.

Sliter, W. V. 1971. Predation on benthic foraminifers. *J. Foram. Res.* 1: 20–29.

Smetacek, V. 1981. The annual cycle of protozooplankton in Kiel Bight. *Mar. Biol.* 63: 1–11.

Smith, H. G. 1973. The ecology of protozoa in chinstrap penguin guano. *Br. Antarct. Surv. Bull.* 35: 33–50.

Sonneborn, T. M. 1957. Breeding systems, reproductive methods, and species problems in Protozoa. Pp. 155–324 in E. Mayr (ed.), *The Species Problem*. AAAS Publ., Washington, D.C.

Sonneborn, T. M. 1959. Kappa and related particles in *Paramecium*. *Adv. Virus Res.* 6: 229–356.

Sonneborn, T. M. 1975. The *Paramecium aurelia* complex of fourteen sibling species. *Trans. Am. Microsc. Soc.* 94: 155–178.

Sørensen, J., Jørgensen, B. B. & Revsbech, N. P. 1979. A comparison of oxygen, nitrate, and sulfate respiration in coastal marine sediments. *Microb. Ecol.* 5: 105–115.

Sorokin, Yu. I. 1965. On the trophic role of chemosynthesis and bacterial biosynthesis in water bodies. *Mem. Ist Ital. Idrobiol.* 18 Suppl.: 187–205.

Sorokin, Yu. I. 1977. The heterotrophic phase of plankton succession in the Japan Sea. *Mar. Biol.* 41: 107–117.

Sorokin, Yu. I. 1978. Description of primary production and the heterotrophic microplankton in the Peruvian upwelling region. *Oceanology* 18: 62–71.

Sorokin, Yu. I. 1981. Microheterotrophic organisms in marine ecosystems. Pp. 293–342 in A. R. Longhurst (ed.), *Analysis of Marine Ecosystems*. Academic Press, London.

Sorokin, Yu. I. & Paveljeva, E. B. 1972. On the quantitative characteristics of the pelagic ecosystem of Dalnee Lake (Kamchatka). *Hydrobiologia* 40: 519–552.

Spero, H. J. 1982. Phagotrophy in *Gymnodinium fungiforme* (Pyrrhophyta): the peduncle as an organelle of ingestion. *J. Phycol.* 18: 356–360.

Spielman, L. A. 1977. Particle capture from low-speed laminar flows. *Ann. Rev. Fluid. Mech.* 9: 297–319.

Spittler, P. 1973. Feeding experiments with tintinnids. *Oikos*, suppl. 15: 128–132.

Steele, J. H. 1974. *The Structure of Marine Ecosystems*. Harvard University Press, Cambridge, Mass.

Stout, J. D. 1980. The role of protozoa in nutrient cycling and energy flow. *Adv. Microb. Ecol.* 4: 1–50.

Stout, J. D. 1984. The protozoan fauna of a seasonally inundated soil under grassland. *Soil. Biol. Biochem.* 16: 121–125.

Stout, J. D., Bamforth, S. S. & Lousier, J. D. 1982. Protozoa. Pp. 1103–1120 in *Methods of Soil Analysis, 2. Chemical and Microbiological Properties*. Agronomy Monograph, 9, ASA-SSSA, Madison, Wisconsin.

Stuart, V., Lucas, M. I. & Newell, R. E. 1981. Heterotrophic utilization of particulate matter from kelp *Laminaria pallida*. *Mar. Ecol. Prog. Ser.* 4: 337–348.

Stump, A. B. 1935. Observations on the feeding of *Difflugia, Pontigulasia* and *Lesquereusia. Biol. Bull.* 69: 136–142.

Sudzuki, M. 1979. Psammobiont Rhizopoda and Actinopoda from marine beaches of Japan. *Acta Protozool.* 18: 293–304.

Swanberg, N. R. 1983. The trophic role of colonial Radiolaria in oligotrophic oceanic enviroments. *Limnol. Oceanogr.* 28: 655–666.

Swanberg, N. R. & Anderson, O. R. 1985. The nutrition of radiolarians: trophic activity of some solitary Spumellaria. *Limnol. Oceanogr.* 30: 646–652.

Taylor, F. J. R. 1982. Symbiosis in marine microplankton. *Ann. Inst. Océanogr., Paris* 58: 61–90.

Taylor, F. J. R., Blackbourn, D. J. & Blackbourn, J. 1971. The red-water ciliate *Mesodinium rubrum* and its "incomplete symbionts": a review including new ultrastructural observations. *J. Fish Res. Bd. Canada* 28: 391–407.

Taylor, G. T. 1982. The role of pelagic heterotrophic protozoa in nutrient cycling: a review. *Ann. Inst. Océanogr., Paris* 58: 227–341.

Taylor, W. D. 1978. Maximum growth rate, size and commonness in a community of bacterivorous ciliates. *Oecologia (Berl.)* 36: 263–272.

Taylor, W. D. 1983. A comparative study of the sessile, filter-feeding ciliates of several small streams. *Hydrobiologia* 98: 125–133.

Tendal, O. 1972. *A monograph of the Xenophyophoria.* Galathea Report 12: 7–102 + plates 1–17. Danish Science Press, Lt., Copenhagen.

Trench, R. K., Pool, R. R., Logan, M. & Engelland, A. 1978. Aspects of the relationship between *Cyanophora paradoxa* and its endosymbiotic cyanelles *Cyanocyta korschikoffiana* (Hall and Claus). I. Growth, ultrastructure, photosynthesis and the obligate nature of the association. *Proc. R. Soc. London.* B, 202: 423–448.

Trinci, A. P. J. & Thurston, C. F. 1976. Transition to a nongrowing state in eukaryotic microorganisms. Pp. 55–79 in T. R. G. Gray and J. R. Postgate (eds.), *The Survival of Vegetative Microbes.* Cambridge University Press, Cambridge.

Tucker, J. B. 1968. Fine structure and function of the pharyngeal basket in the ciliate *Nassula. J. Cell Sci.* 3: 493–514.

Van Bruggen, J. J. A., Stumm, C. K. & Vogels, G. D. 1983. Symbiosis of methanogenic bacteria and sapropelic protozoa. *Arch. Mikrobiol.* 136: 89–96.

Van Houten, J., Hauser, D. C. R. & Levandowsky, M. 1981. Chemosensory behavior in protozoa. Pp. 67–124 in M. Levandowsky & S. H. Hutner (eds.), *Biochemistry and Physiology of Protozoa.* 2nd ed., vol. 4. Academic Press, New York.

Veldkamp, H. & Jannasch, H. W. 1972. Mixed culture studies with the chemostat. *J. Appl. Chem. Biotechnol.* 22: 105–123.

Verity, P. G. 1985. Grazing, respiration, excretion and growth rates of tintinnids. *Limnol. Oceanogr.* 30: 1268–1282.

Vogels, G. D., Hoppe, W. F. & Stumm, C. K. 1980. Association of methanogenic bacteria with rumen ciliates. *Appl. Environ. Microb.* 40: 608–612.

Volz, P. 1972. Über Trockenresistenz und Wiederaufleben der Mikrofauna (inbesondere Protozoen) des Waldbodens. *Pedobiologia* 12: 156–166.

Watson, J. M. 1946. The bionomics of coprophilic protozoa. *Biol. Rev.* 21: 121–139.

Webb, M. G. 1956. An ecological study of brackish water ciliates. *J. Anim. Ecol.* 25: 148–175.

Webb, M. G. 1961. The effects of thermal stratification on the distribution of benthic protozoa in Esthwaite water. *J. Anim. Ecol.* 30: 137–151.

Wesenberg-Lund, C. 1925. Contributions to the biology of *Zoothamnium geniculatum* Ayrton. *D. Kgl. Danske Vidensk. Selsk. Skrifter, Naturvidensk. og Matematik, Afd.* 8, *række* X. 1: 1–53 + plates I–XIV.

Wessenberg, H. & Antipa, G. A. 1970. Capture and ingestion of *Paramecium* by *Didinium nasutum*. *J. Protozool.* 17: 250–270.

Wheeler, P. A. 1983. Phytoplankton nitrogen metabolism. Pp. 309–346 in E. J. Carpenter & D. G. Capone (eds.), *Nitrogen in the Marine Environment*. Academic Press, New York.

Wilbert, N. & Kahan, D. 1981. Ciliates of Solar Lake on the Red Sea shore. *Arch. Protistenk.* 124: 70–95.

Wille, J. J. Jr., Weidner, E. & Steffens, W. L. 1981. Intranuclear parasitism of the ciliate *Euplotes* by a trypanosomatid flagellate. *J. Protozool.* 28: 223–227.

Williams, G. C. 1975. *Sex and Evolution*. Princeton University Press, Princeton.

Williams, P. J. leB. 1981. Incorporation of microheterotrophic processes into the classical paradigm of the planktonic food web. *Kieler Meeresforsch., Sonderheft* 5: 1–28.

Winkler, R. H. & Corliss, J. O. 1965. Notes on the rarely described, green colonial protozoon *Ophrydium versatile* (O.F.M.) (Ciliophora, Peritrichida). *Trans. Am. Microsc. Soc.* 84: 127–137.

Zeitzschel, B. 1982. Zoogeography of pelagic marine protozoa. *Ann. Inst. Océanogr., Paris* 58: 91–116.

Zeuthen, E. 1953. Oxygen uptake as related to body size in organisms. *Quart. Rev. Biol.* 28: 1–12.

# Index

Numbers in italics refer to figures

*Acanthamoeba* 161
Acantharia 79,114
*Acaryophrya* *139*,140
*Acineta* 48,*164*,165
*Actinobolina* 52,*139*,140
*Actinomonas* 108
*Actinophrys* 23
*Actinosphaerium* *23*,49,142
Activated sludge 150
Alkaline lakes 74
*Allogromia* 124–125
*Ammonia* 127
*Amoeba* *23*,142
Amoebae, in soils 154–155,158–159
*Amphidinium* 120
*Amphisella* 155
*Arcella* *143*
*Archnula* 154
*Aspidisca* 120,133,142
*Astrorhiza* *124*
Autogamy 7,69–71
Axopodia 22,24

Balanced growth 53–54
*Balantidium* 162
*Blepharisma* 29,43,67,120,127,129,142
*Bursaria* 36,142

*Caenomorpha* 85,121,128,143
*Centropyxis* *143*
*Cercobodo* 142,155
*Chaos* 13,98
*Carchesium* 148
Chemosensory behavior 28
*Chilodonella* 148

Chitinozoa 4
*Chlamydodon* 120,127,129
Chloroplast symbiosis 81
Choanoflagellates:
   feeding mechanisms 41,43
   in plankton 106,108,137
Chrysomonads, in plankton 106–108
Cilia:
   function of 17–21
   thigmotactic 22
Ciliates:
   feeding mechanisms 43–45,46–49,52
   in limnic plankton 137–140
   in limnic sediments 142–145
   in marine plankton 111–114
   in marine sediments 119–123,125,
    127–129
   in soils 153,156–157,158–160
*Ciliophrys* 108
Clearance 33–35
*Climacostomum* 82,127
Codonella 137
*Codosiga* *138*
Coexistence of competing species 92–97
*Coleps* 49,127,140,142
*Colpidium* 36,43,44,142,149
*Colpoda* 44,63–64,67,153,*155*,156
*Condylostoma* 82,99,127
*Conidiophrys* *164*,165
Conjugation 7,69–71
*Cothurnia* 133,*164*
*Cristigera* 120,129
Cyanella 77,78
*Cyanophora* 78
*Cyclidium*
   39,73,113,127,129,131,140,149

195

Cysts 63–64,153

Desiccation, ecological factor for soil
    protozoa 153
*Desmarella 107*
Detritus, protozoan communities of 129–
    133
*Diaphanoeca* 11,*11,107*
*Didinium* 36,46,64,92,95,140,142
*Difflugia* 140,*143*
*Dileptus* 30,46,142
Dinoflagellates, in plankton 113–114
*Diophrys* 120,127
*Discocephalus* 120
Dispersal 99
*Dysteria* 133

Ectosymbionts, bacterial 84
*Elphidium 51,*81,127
Endemic distribution 98
Endosymbionts:
    bacterial 83–85
    photosynthetic 78–83
*Entosiphon* 46
*Epistylis* 148
*Erythropsis* 10,*11,*22
*Euglena* 28,29,74
*Euglypha* 22
*Euplotes* 10,*10,*19,*20,*30,39,43,67–
    68,70–71,82,127,133,*139,*140
*Eutintinnus 94,112*
Extreme environments 74–75

*Favella 94,112*
Filopodia 22
Flagella, function of 17–21
Flagellates:
    feeding mechanisms 41–43,45–46
    in plankton 106–111,136–137
    in sediments 123,125,127,145
    in soils 155–156,159
Food chains, models of 87–89
Food retention, mechanisms of 35–41
Foraminifera:
    in plankton 50–51,114
    in sediments 52,79–81,123–125,127
*Frontonia* 47,*48,*82,121,127,129,*146*

Gamones 28
*Geleia* 120
Geotaxis 29,30–31
*Glaucoma* 44,142,149
*Grossglockneria 155,*156
Growth efficiency 54–55
Growth rate 57–59
*Gruberia* 129
*Gymnodinium* 46
Gymnodinioides 165
*Gyrodinium* 46,113

*Halteria 42,*43,137
*Hartmannula* 133
*Hastigerina* 50
*Helicostomella 94*
Helioflagellates, in plankton 108
*Hemiophrys* 133
*Hexamita* 61
Histophagy 49,163
*Holosticha* 127,133,142,*155*
*Homalozoon* 46
*Hyalodiscus* 73
Hydrogenosomes 62
*Hypocoma 164,*165

*Ichthyophtirius* 67,163
Interstitial fauna 119–123
Island biogeography 99–100

*Kentrophorus* 85,120
*Keronopsis* 133,142

*Laboea* 111,*112*
*Lacrymaria* 46,127,142
*Lembadion* 47,68,*139*
*Leptomyxa* 154
*Litonotus* 36,46,127,133,142
Lobopodia 22
*Lohmanniella* 111,*112*
*Loxodes* 10,*11,*28,29,31,62,74,
    94,145,*146*
*Loxophyllum* 46,120,127

*Marsipella 124*
*Mastigamoeba* 155
Mastigonemes 17
Mating types 70
Mechanoreception 30
*Megamoebamyxa 126*
Membranelles, function of 19,43–45
*Mesodinium* 82–83,113
Metabolism:
    anaerobic 61–62
    rate of 57–59
*Metopus* 85,121,128,143,162
Microbial loop 103–104
*Microthorax 155*
*Monas* 106,*107*
*Monosiga 107*
Motility 16–24
    control of 24
    energetic costs of 55
    sliding 22
Müller vesicle 10,*11*
*Myelostoma* 121,128
Myonemes 22,24

*Naegleria* 64–65,154,161
*Nassula* 47,142
*Nebela 143*

*Noctiluca* 22,46,*112*,114
*Nonion* 81
*Nummulites* 79
Nutrition, by dissolved organics 39–41

*Ochromonas* 45,*45*,60,106
*Ophrydium* 143
*Ophryoglena* 49,*50*,67,142
Osmotic pressure, as ecological factor
 73–74
*Oxyrrhis* 46,113

*Parablepharisma 80*, 121,127
*Paradileptus* 140
*Paramecium* 8,9,28,29,30,36,43,60,70–
 71,*80*,81–82,83–84,92,93,95,129,142
*Paraphysomonas* 54,106,*107*,108
*Paraspathidium* 129
Patchiness, environmental, 90–92
Patchy resources, adaptations to 59–
 61,63–68,90–92,95,97
*Pelomyxa* 13,*23*,85,142
*Peranema* 46
Percolating filters 150
*Peridinium* 78
*Peritromus* 129
pH, as an ecological factor 74
Phagocytosis 32–33
Photoreception 29–30
Pigments, of ciliates 29
Pinocytosis 41
*Plagiopyla* 121,128,143
Plankton communities, spatial structure
 of 105
*Pleuromonas* 45,*45*
*Pleuronema* 120
Pollution 149–150
*Polykrikos* 5,*6*
Prey-predator oscillations 91–92
*Prorodon* 49,127,129
*Prospathidium 155*
*Psammosphaera 124*
*Pseudobodo* 108
Pseudopodia, function of 22–24
*Pseudoprorodon* 120,*121*
*Pteridomonas* 108

Radiolaria 52,78–79,114
*Remanella* 93,*94*,120,*121*
Remineralization, role of protozoa 88–
 89,115,132–133
Responses:
 kinetic 26–28
 phobic 26
 taxic 25,27
 transient 26–28
Reticulopodia 22
Reynolds number 16

Salinity, as ecological factor 73–74,126–
 127
*Saprodinium* 121,128
Sarcodines, feeding mechanisms 49,52
Sediments:
 chemical properties of 116–118
 mechanical properties of 118–119
Sexuality 6–7,68–72
Size range, of protozoa 11–14
*Sonderia 48,80*,84,121,128
*Sorogena* 64,65,156
Spasmoneme 24
*Spathidium* 142,156
Species, concept of 7–9
*Spirostomum*13,127,*144*,145,*146*
*Stannophyllum 124*
Starvation, responses to 59–61,64
*Stenosomella 112*
*Stentor* 13,39,43,82,127,142
*Sticholonche* 22
*Strombidium* 48,82,111,120,127,137
*Stylonychia* 142
Sulfureta 127–129
Symbiosis, definition of 76–77

Temperature, as an ecological factor 75–
 76
Testate amoebae, in soils 154,159
*Tetrahymena* 9,28,39,40,43,49,60,65–
 67,70–71,98,142,161
*Tetramitus* 64–65
*Thecamoeba* 155
*Tintinnidium* 137,*138*
*Tintinnopsis 112*,137
*Tontonia* 22,111
Toxicysts 46
*Trachelius* 140,142
*Tracheloraphis* 120,*121*
*Trepomonas* 61
*Trichodina* 163
*Trochammina* 127
*Trochilia* 129,133

*Urceolaria* 163
*Uronema* 113,116,131,140
*Uronychia* 47

*Vahlkampfia* 154
*Vanella* 148
*Volvox* 5,6
*Vorticella* 22,39,133,140,*144*,148

*Xenophyophoria* 13,*124*,125–126

*Zoochlorella* 77
*Zoothamnium* 6,*6*,133,*164*
*Zooxanthella* 77